Do You Speak Bee?

The Incredible Story of Lives Inside the Hives

by
Albert B. Chubak

One Printers Way
Altona, MB R0G 0B0
Canada

www.friesenpress.com

Copyright © 2024 by Albert B. Chubak
First Edition — 2024

Illustrations by Huw Evans

All graphic/drawing/artwork illustrations contained in this book are the work of Huw Evans unless otherwise stated and may not be reproduced or copied in any form without the express permission of the author and illustrator.

All rights reserved.

No part of this publication may be reproduced in any form, or by any means, electronic or mechanical, including photocopying, recording, or any information browsing, storage, or retrieval system, without permission in writing from FriesenPress.

ISBN
978-1-03-830037-9 (Hardcover)
978-1-03-830036-2 (Paperback)
978-1-03-830038-6 (eBook)

1. NATURE, ANIMALS, INSECTS & SPIDERS

Distributed to the trade by The Ingram Book Company

Dewey M. Caron

There are a good number of bee books but none quite like *Do you Speak Bee?* In addition to older teens, beekeepers and individuals of any age interested in honey bee social life and colony organization will be entertained and informed with this new book by Albert Chubak. It is a delightful, informative book, perfect in everyone's bee library.

Bees inform each other about their societal needs mainly via "dancing", repeated movements that encode flower and homesite information and chemical emissions of amazing complexity. There are hive bees capable of attending to tasks that need attention. This language is decoded in *Do you Speak Bee* illustrations and text that accurately and interestingly invite us to understand the city of bees and how their society depends on communication to accomplish everyday tasks. Do You Speak Bee is a fascinating illustrative representation of the intricate bee world. Welcome and enjoy bee speak.

Richard E. L. (Dick) Rogers

Graphics amazing!

Comprehensive in content, and well organized.

Interesting subject and approach that draws in the reader.

A new way of thinking about bee tasks and roles that is lighthearted and sometimes humorous.

This will be an interesting read for anyone who wants to learn more about honey bees.

Table of Contents

Preface	ix
Introduction	**1**
Do Bees Wear Hats?	1
Is it Proper to Write Honeybee or Honey Bee?	1
Bee Terms in Singular vs. Plural	2
Imperial and Metric	2
Family History of a Honey Bee	2
Pollinators Come in a Variety of Evolutionary Costumes	4
A Honey Bee Colony is an Insect City with Age-based Jobs that are Equally Important	4
The Boutique of Fragrances in a Hive Tells a Story	5
Hormones, Pheromones, Kairomones, Allomones and Synomones are all Messengers	6
Glands in Honey Bees are the Honey Bee's Resource Tool Bag	6
Built in Honey Bee Equipment	7
Bee Development and Age-based Duties	7
Cast(e) of Characters in a Honey Bee Colony	8
The Honey Bee Colony	**9**
Life Begins in the Brood Nest	9
Queen Bee	11
Worker Bee	13
Drone Bee	14
The World of the Queen Bee	**15**
Queen Bee	15
Adopted Bee	17
Fighting Virgin Queen Sisters	18
Royal Executioner	19
Royal Excrement Bee	20
Swarming Bee	21
Trojan Horse Bee	22
Vibrating the Queen Bee	23

The Amazing Worker Bee:	
Activity in the Brood Nest Nursery	**25**
Nurse Bee	25
Attendant to Queen	27
Bee Bread Pounder Bee	28
Brood Heater Bee	29
Cell Capper	30
Childcare Nursery Inspector	31
Cell Size Bee Inspector	32
Laying Worker Bee	33
Propolis Cell Polisher	34
Royal Jelly Maker	35
Wax Chewer	36
Worker Bees Jobs Inside the Hive but Outside of the Brood Nest	**37**
House Bee	37
Biomonitoring Bee	38
Blower Bee	39
Congress Bee	40
Cook Chef	41
Disaster Cleanup	42
Disease Inspector Bee	43
Dock Worker	44
Engineer Bee	46
Fanning Bee	47
Grooming Bee	48
H-VAC Bee	49
Line Dancer Washboarding	50
Listening Bee	52
Master Environmentalist and Herbalist Bee	53
Mobile Food Services	54
Propolis Applicator Bee	56
Recycling Bee	57
Relocation Services	58

Shaker Trembling Bee	59
Sleeping Bee	60
Traffic Control	61
Thermal Defense Bee	62
Unemployed Bee	63
Waggle Dancer	64
Winter Bee	65

Field Bees — 67
- Field Bee — 67
- Absconding Bee — 69
- Air Defense Bee — 70
- Bee-Lining Bee — 71
- Cleansing Bee — 72
- Drifter Bee — 73
- Flower Pollen Dancer — 74
- Foraging Bee — 75
- Garbage Removal — 76
- Guard Bee — 77
- Orientation Bee — 78
- Packaged Bee — 79
- Robber Bee — 80
- Security Enforcer Bee — 81
- Security Examination Bee — 82
- Senior Bee — 83
- Scavenger Bee — 84
- Scout Bee — 85
- Undertaker — 86
- Veteran Bee — 87
- Water Bee — 88

Drone: The Mating Specialist — 89
- Drone — 89
- Band of Brother-Husbands — 91
- Drone Foster Family — 92
- Drone Fraternity — 93
- Drone Grooming — 94
- Drone Race Track — 95
- Food Beggar Drone — 96
- Freezing Drone — 97
- Predator Fodder — 98

Activities & Famous Bees — 99
- City Bee — 100
- Fossilized Bee — 101
- Garden of Eden Bee — 102
- Pilgrim Bee — 104
- Weaponized Bee — 106
- White House Bee — 107
- Wild Feral Bee — 108

Closing Remarks — 109

Dedication

I dedicate this book to:

My mom and sisters, who have always believed in me.

My brothers, who have always been behind me.

My children, Abi, Anna, and AJ.

My dear wife Camellea.

Daniel, my dear homeless friend, who offered countless hours of friendship and support.

My friend Bart, for being a continuous resource of information.

Huw Evans, my amazing artist, adviser, and sounding board.

My bee friends Barry, Charlie, Dana (Ashe), Danielle, David, Joe, Marlene, Michael, Paul, Sam, Veldon, and Walter for their friendship, mentoring, and insights.

A huge thanks to Dewey Caron, Richard (Dick) Rogers, Tammy Horn Potter, and Della Konkin for their content support.

Kim Flottum, for his unfailing support.

The American Beekeeping Federation, Eastern Apicultural Society, Western Apicultural Society, Heartland Apicultural Society, Wasatch Beekeeping Association, Utah County Beekeepers Association, Utah State Beekeepers Association, Wyoming Bee College, Santa Barbara Beekeeping Association, Jordanian Beekeepers Association, Saskatoon Bee Club, and Bee Culture Magazine for the vital parts they each played in my development as a beekeeper, educator, and writer.

My beekeeper friends all over the globe.

All those who love creativity, bees, and the amazing natural world we live in.

Preface

Years ago, while visiting with my sister, I discovered my brother gave her a frame of honeycomb to eat. As I produced honeycomb for sale at that time, I was disappointed I had not given her this amazing, sweet treat first. A bit of sibling rivalry showing here. When I asked her how it was, the response was, "Very juicy. It was running down my face. And later that same week bees began showing up in the house." Her statement was odd, as honey is not juicy, nor would honey bees appear indoors from a capped honey frame. The juicy, and bees flying indoors, suggests the frame was incubating capped larvae and not honey at all. This clearly shows why knowing what is happening in a beehive, and on a frame, is vitally important to both beekeeper and honey connoisseur.

David Bench opened my eyes to the incredible world that existed in a hive. As we caught swarms and performed removals from homes, we saw amazing adaptation by the bees to where they chose to live. One queen specifically, we named the Grim Reaper. The five-foot swarm she was in, resembled this character with a hood, sickle, and tall body as it hung on the tree. Some colonies were so amazing, we chose to graft queens from them, replicating them for other colonies.

Instead of the large box hives, a mating box for the queens to be raised in was created. They were quarter the size of a medium Langstroth box, with new smaller 6-inch square frames. Dave grafted daughter queens from this original Grim Reaper queen. As the season ended, we ended up with extra queens. These were saved in stacked mating boxes and placed inside a basement window for the winter. Prior to spring, an inspection showed the multiple sugar water fed colonies were strong and growing. This led to an expansion of the boxes, and still later to multiple splits to create grand-daughter queens from the original Grim Reaper queen. This exciting new way to expand colonies led to the manufacturing these smaller hives and introducing beekeepers to this simple novel way to start and grow a colony. The work with David led to the creation of a hive called the Mini Urban Beehive (MUB). This hive helped tutor me on the complex world that exists inside each hive. This MUB hive is about a quarter size of a traditional white painted boxed hive seen in farmer's fields. It can expand from a single box to four or more boxes. This hive is an incubator, where a queen can be raised, or a tiny colony can start. In nature for millions of years, bees lived in hollow cavities inside trees. It is a vertical hive, where boxes can be added as the colony grows. While a colony is raising their queen, and beginning to grow, the colony is usually docile. This period is ideally the best time to investigate and learn what and why things are happening. For example, when a colony goes queenless, within hours the young nurse bees know the queen is missing and start adding more royal jelly to selected young larvae. This continuous feeding of female larvae generates queens to be started. There could be one queen created or there could be multiple queens created all at once. This begins a cycle that can easily be followed. Participating in this creation is exciting, like having a child but on a smaller scale. Resources can be offered to help this growing colony, but the interaction continues, as protection is also needed. During this expansion and growth, incredible lessons can be learned. Learning life cycles, food requirements, the progression of bee duties, differences of nectar and honey, pollen and bee bread, capped honey, capped larvae, and so much more can be seen if the time is taken to look. While this colony is small it is allowing for an intimate relationship with the learning beekeeper. Once the colony has grown, boxes can be added, or the multiple boxes can be separated and the one without a queen goes through the cycle of raising a new queen again. The amazing thing with the little hive, was that while bee population was small and devoted to either creating a queen or building their colony, they were quite docile. During this small window of opportunity, great learning began to occur. Questions began to flow, why were they doing this, or when would this happen, or what do they need to do better? Now the incredible details of the innate workings of the colony were more visible than ever.

This led to a decision of raising bees instead of focusing on honey. It occurred to me that the hive was first an incubator and second a honey production plant. Since then, until now, my focus has been on creating bees and teaching the other side of beekeeping. Not honey, but how bees grow and what they are doing. The MUB hive started circulating locally, then across the U.S., and finally they were sent around the world. Then in 2014, I was honored to receive Utah's Best of State award for Most Innovative Company in Utah.

In 2017, I created a publication on this MUB hive and as a direct result, began writing for new beekeepers in the American Beekeeping Federation in their quarterly publication. Four of my articles were published in the Bee Culture magazine. This same year, the Western Apicultural Society awarded me their P.F. Thurber award for Inventiveness. In 2019, a series of reference letters were given with an application to the Eastern Apicultural Society (EAS) to be admitted into their Master Beekeeping examination. Still, as of 2023, more work and study must occur.

Understanding of the honey bee has become an unquenching focus. In 2018, I penned another article called, "Reading a Frame." It was an insightful article that helped readers understand what the bees were saying on a bee frame. At Apimondia 2019 in Montreal, the idea of authoring a book focused on what jobs bees have and how they communicate. Initially only 20 bee roles could be identified. Later the list grew to almost a hundred, and now extends to about 150 different bee activities and roles. About 50 images will be left for a later book.

Huw Evans was contracted to draw the images. He also had to learn about the bee, its biology, variations between the drone, queen, and worker bee. His creativity in the creation of each of the images is incredible.

This work has merged playful art, biology, beekeeping, and historical events surrounding the honey bee into a wonderful educational book on bees from children to adults.

Introduction

Do Bees Wear Hats?

In a sense, they do. From crowns to cowboy hats and hard hats to hair nets, all styles of head coverings are seen as symbols standing for a person's culture, status, and especially their occupation. A statement like "What hat is being worn today?" signifies the duty the person is performing, whether they are wearing a hat or not. Given their symbolic or figurative meaning, bees, especially the female workers, wear hats throughout their short working lives. A bee's lifetime of chores first begins when it emerges after its transformation through larval and pupal stages into this busy honey bee world. As a newborn colony citizen, bees know instinctively to clean and polish their vacated birth cell with propolis, a bee's disinfectant. In this way, a bee's first hat is that of a house cleaner, but that is only the first hat they wear.

The following pages illustrate, in a fanciful cartoon way, the amazing variety of hats bees wear and what they are doing when it is said "As busy as a bee."

Is it Proper to Write Honeybee or Honey Bee?

The Merriam-Webster dictionary shows both Honeybee, and Honey Bee, as acceptable writing variants. People who study insects are called **entomologists**. Entomologists have a cardinal rule when using insect names. If the insect is what the name claims, write the two words separately, otherwise join the two words together. For example, a house fly and a horse fly are both part of the fly order of insects or taxonomic grouping (scientific classification), so the two words are separate. A butterfly and dragonfly are not flies and are in different taxonomic orders, so their names are joined. A honey bee is a bee and belongs to the order of insects having the bee family, so the name is written as two separate words. In other words, a honey bee is a bee that produces honey. Not all bees are honey producers. As a result, throughout this book, honey bee as two separate words will be used.

Bee Terms in Singular vs. Plural

Larva – Larvae
Pupa – Pupae
Egg – Eggs
Bee – Bees
Insect – Insects

If the singular term ends with the letter "a," making it plural requires adding the letter "e." An example for singular larva in a sentence: The larva was wiggling in a cup. The plural version of that same sentence adds an "e" to larva to become: The larvae were wiggling in a cup.

Imperial and Metric

In the following pages, temperature, measurements, and weights are noted in both, the Imperial, and Metric system of measurements. In beekeeping, tiny measurements are preferred in Metric opposed to Imperial, such as in measuring the diameter of the honey bees' wax cells. To use fractions to describe sizes less than a half inch is avoided, and millimeters are used. Simple conversions are as follows: Freezing point of water is 32°F and is 0°C. One inch is 25.4 mm. One millimeter is close to the same as 0.04 inches. One pound is the same as 0.45 kilograms. One kilogram is close to the same as 2.2lbs. Conversion calculators exist online for further clarity.

Family History of a Honey Bee

The media discusses the decline of bee populations in an overly broad fashion. There are more than 20,000 species of bees in the world. Bees are threatened globally, but in North America, only one bee is on the endangered list, the Rusty Patched Bumble Bee. In an ever-changing world, insects may struggle and need external help. The same is true of the nearly 10 million other species of organisms. Life on Earth is dynamic and constantly undergoing changes in diversity and abundance. Humans have an impact on the speed and intensity of the impact. Not including plants and fungi and a couple of other groups of living things, Earth is covered with an amazing number of organisms that belong to the kingdom Animalia. You are a member of this kingdom. You are an evolutionary cousin to the honey bee, here is how. Both you and the honey bee are part of the Kingdom of Animals. This is the family lineage and history of the bee we call the honey bee.

Kingdom: Animalia or Animals
There are five to six kingdoms. Animals, plants, and fungi are the three major kingdoms.

Phylum: Arthropods
Joint-footed invertebrates have segmented bodies with segmented limbs and chitin for an outer protective shell. This is the most common type of animal on Earth.

Subphylum: Uniramians
This includes Hexapod insects, Myriapoda centipedes and millipedes, and velvet worms. They have single branch appendages, one pair of antennae, and two pairs of mouthparts (mandibles and maxillae).

Class: Insects
These have three body sections: head, thorax, and abdomen. They have a pair of antennae, three pairs of legs, and up to two pairs of wings.

Order: Hymenoptera
The word Hymenoptera is made up of two Greek words. The first part of the word is hymen, which means membrane. The second is -ptera, which translates in Greek to be wing. Put them together to mean wing-membrane. These insects have a tubular or lance-like structure at the end of the abdomen that is used to sting, lay eggs, and/or inject eggs into a host. Honey bees no longer use the ovipositor as an egg-laying tool; now it is solely used as a stinger for defense. This group includes ants, wasps, and bees.

Family: Apidae
There are more than 20,000+ species of bees in the superfamily of bees called Apoidea. Within this grouping is the family of bees called Apidae, which has 6,000+ species globally. Bees in the Apidae family have a long tongue consisting of intricate parts, which together are called the proboscis. Apidae bees are vegetarians. Various species in this family have evolved different adaptations for carrying pollen. While honey bees (Apis species) are well-known for their pollen baskets (corbiculae) on the hind legs, other bees within the family may have alternative structures or methods from transporting pollen. These variations include Scopa hairs, Pollen combs, Pollen brushes, and Pollen-carrying structures on the body. These adaptations reflect the ecological niches and foraging behaviours of the various species.

Genus: Apis
This is a honey-producing bee. Trigona Prisca was a stingless honey bee from the Cretaceous period, 96–74 million years ago. The oldest honey bee fossil in the Americas was found in the Nevada shale, dating back 11 million years. There are eight species of honey bees alive today. However, there are currently 43 subspecies of these eight species:

- *Apis florea* Red Dwarf Asia,
- *Apis dorsata* Giant Asia Nepal,
- *Apis cerana* Asia,
- *Apis laboriosa* Himalayan Giant,
- *Apis koschevnikovi* Malaysia and Indonesia,
- *Apis andreniformis* Black Dwarf tropical Asia,
- *Apis mellifera* European,
- *Apis nigrocincta* Philippines.

Types: Mellifera

Mellifera means bearing honey. This group is used for honey production worldwide. This is the common Western honey bee, also called the European honey bee. There are seven races of European / Western Honey Bees:

- *Carnica* or *Carniolan* – Eastern Europe
- *Caucasica* or *Caucasian* - Central Caucasus in Georgia
- *Iberiensis* or Iberian or Spanish or Gibraltar – Iberian Peninsula in Spain and Portugal
- *Mellifera* or European Dark or German – Iberian Peninsula to Russia
- *Liguistica* or Italian – Italy
- *Linnaeus* or *Capensis* – South Africa
- *Scutellata* or East African Lowland – South Africa

(Africanized, Buckfast, Cordovan, Starline, Midnite, Minnesota Hygienic, and Saskatraz bees are mixed Mellifera race variations.)

Pollinators Come in a Variety of Evolutionary Costumes

Using a term like "bees" is an overly broad way to refer to honey bees. There are other types of wild bees besides honey bees. Those who care for other bee species are considered beekeepers too. Bumble bees live in nests consisting of about 40 bees. Bumble bees can be managed by humans and used inside greenhouses for controlled pollination. Solitary bees include alfalfa leaf-cutter bees, blue orchard mason bees, carpenter bees, miner bees, sweat bees, and others. Solitary bees are pollination specialists and do not create surplus honey. There are also varieties of stingless bees. Stingless bees live in colonies and produce harvestable honey. All bees are united with honey bees in pollination. Bees pollinate very well, but without all of them, plants may not bear nuts, fruit, or vegetables. There are plants that can be pollinated without bees. Pollinators come in a variety of evolutionary costumes. Other insects can pollinate blossoms, like butterflies, moths, ants, flies, and wasps. Animals like bats and birds pollinate, as do wind and rain. Pollination is a dynamic term backed by Mother Nature's greatness, with many natural pollinator redundancies.

A Honey Bee Colony is an Insect City with Age-based Jobs that are Equally Important

Honey bees live in highly interactive social groups called **colonies**. These social communities are like a city or town, with multiple contributors performing age-based jobs. In any global city, a mayor is not enough; there needs to be city workers, storage facilities, transportation, farmers producing and supplying food, health care resources, places for the young to be cared for, and more. In this way, honey bees and people are very much the same. If one area in the city or colony is neglected or lacking, the system is failing and is unhealthy. For a colony to be healthy, all jobs in the colony need to be staffed, and they need to be free of disease and low in invasive parasitic threats.

Introduction

The Boutique of Fragrances in a Hive Tells a Story

Bees communicate mostly with scents called pheromones. Everything has a scent that communicates a message to a bee.

Scents in a hive send messages, like the varied scents that occur each day in a home. Smells and aromas constantly send messages to those nearby. These daily scents vary from baking in the kitchen to freshly cleaned laundry, a spring-scented breeze blowing through the window, smells related to illness, paint, exhaust fumes in the garage, and puffs of smelly gas from the restroom. Each house scent has a meaning for those who smell it and creates a call to action. Bees have similar messages conveyed through pheromones. A variety of scents exist at any one time in a hive. Scents created by honey bees that function as signals are called **pheromones**. Pheromones are **primers (major),** and others are **receivers (minor)**. Primer pheromones control the entire colony and encourage growth from one stage to another. These primer pheromones last for days. Receiver pheromones are immediate pheromones used for a local issue, like an alarm pheromone. Alarm pheromones only stir up bees in the vicinity, and they dissipate in minutes to hours. Each pheromone smell has a message attached to it for bees to decipher. A queen bee has a pheromone she emits called the **Essence of the Queen**. Workers have pheromones they share. One worker pheromone can alert other bees to danger and is called an **alarm pheromone**. The nursery with eggs and larvae sends a pheromone throughout the hive, signaling bees to produce wax, secrete royal jelly, and forage for pollen. Imagine bees having a progressive graduation pheromone that leads them to all the future jobs they undertake throughout their short lives. All pheromones communicate specific messages. Scents are a bee's way of communicating and interacting with the environment around them. Even for the reader, scents share a story; just close your eyes, smell the surroundings, and see what they tell you.

Hormones, Pheromones, Kairomones, Allomones and Synomones are all Messengers

Hormones send chemical messages within the human body. Hormone messages can signal growth, how the body turns food into energy, reproduction, mood, and more. With honey bees, pheromones are the primary message senders. But still, there are other message carriers such as **kairomones**, **allomones**, and **synomones**. Kairomones are messages released by bees that sadly attract harmful predators. An example of kairomones in honey bees is a chemical released by drones during the larval and pupal stages that attracts the varroa mite as a predator. As a predator, the mite feeds on the defenseless drone. Allomones are messages released by honey bees that signal help from others but do not help themselves. An example of an allomone in honey bees is the scent released when a bee stings. This scent singles out other defense bees for action but does not help the original stinging bee. Synomones are messages shared between two groups that help them both. An example of synomones is found in the union of the honey bee and the blossom the bees visit. Both the bee and the flower gain from the partnership.

Glands in Honey Bees are the Honey Bee's Resource Tool Bag

Dufour gland—identifies eggs laid by a queen from eggs laid by a worker bee.

Koschevnikov gland (sting gland)—an alarm pheromone in workers. Also responsible for the bawling bees, as it shows a loss of queen signal. Recruited workers.

Hypopharyngeal gland—found in the head of the bee. Produces Royal Jelly

The mandibular glands of a queen—creates the queen's essence, which unifies the colony.

Mandibular glands of a worker—creates an alarm pheromone.

Nasonov gland (scent)—found in worker bees and used to mark the hive entrance. Also used during swarming, selecting new queens, and recruiting workers.

Salivary gland—there are two pairs of salivary glands, one in the bee's head and the other in the thorax. These glands help with the digestion of food.

Tergal glands—found underneath the abdomen and handles uniting the queen's attendants. Adds to the queen's mandibular pheromone.

Tarsal glands (Arnhart) (footprint substance)—this is an oily substance secreted from the arolium pad on the foot. Also called footprint pheromones.

Stinger venom— is the largest of the four sting glands (Venom, Dufour, Koschevnikov, and Nasonov). Releases histamine into the victim.

Wax gland—is on the underneath of the abdomen of worker bees 12–16 days of age. Young bees do not produce wax.

Built in Honey Bee Equipment

Chitin—the exoskeleton of the bee. It serves to protect the bee, like armor.

Tarsal claw and arolium pads—used in harmony to walk on a variety of surfaces.

Antennae—used for touch and smell.

Spiracles—multiple openings on the side of the abdomen that allow the bee to breathe.

Dual wings—a bee has two pairs of wings, two on each side. Hamuli hooks connect a pair of wings while in flight.

Compound eyes—thousands of eyes that allow the bee to see a 360-degree view.

Ocelli/Ocellus—three eyes in a triangle above the compound eyes that gauge light intensity.

Mandibles—these are the pairs of jaws of the bee. Used for carrying, tearing, and biting.

Stinger—the defensive weapon in female bees.

Bee Development and Age-based Duties

Over the course of a bee's lifetime, the worker honey bee engages in progressive age-based activities. These activities can be broadly divided into multiple stages: egg, larval, pre-adult, and adult.

- Egg for 3 days
- Larva for 6 days
- Pupa for 10 days
- Adult bee on day 24 and works for another 45 days

A worker bee egg is laid by a mated queen in the brood nest. After three days, the egg hatches into a larva. Adult nurse bees feed the larva royal jelly. The larva grows rapidly from the size of a grain of rice to a plump, pearly-white-looking pupa. After six days, the larva swims in royal jelly in an open cell, and the larva becomes a large, plump, white pupa. Once the pupa fills the cell, the workers cap the cell, and the pupa stretches out the length of the cell. The developing bee undergoes a metamorphosis, transforming into an adult bee during its 12 days of captivity. The pupal stage lasts for 7½ days for a queen, 12 days for a worker bee, and 14 days for a drone. When the adult bee emerges, it grooms, and its exoskeleton hardens. This exoskeleton is called **chitin**. The new adult bee is ready to join the busy colony in the brood nest. The first job in the brood nest is cleaning its birth cell. Nurse bees are also responsible for secreting royal jelly, cleaning cells, caring for the queen, and feeding the young.

As an adult bee, it progresses from duty to duty effortlessly.

- Day 1–5: worker bees clean and polish cells and cap cells.
- Day 6–12: eat pollen to generate royal jelly to feed brood and queen, tend to the queen, perform hive ventilation, and perform regular queen medical checkups.
- Day 12–17: worker bees become house bees, keeping all the needs of the house in order, transporting resources, dehydrating nectar, producing wax, building combs, and removing dead bees.
- Day 18–21: worker bees perform guarding duties and orientation flights.
- Day 22 on, worker bees forage and function as scouts.

Cast(e) of Characters in a Honey Bee Colony

Honey bees are not all created equal, but all are equally vital. In a colony, bees naturally undertake specific age-based, progressive social duties throughout their lifetime. Initially, all female eggs are the same; nothing is different from a queen egg to a worker bee egg. After the female larvae hatch from the eggs, they are still the identical. On day three, nursery bees change the diet for female honey bees to a honey-based diet. All queen larvae continue to be fed royal jelly. This change sets the stage for female bees to be locked into either a worker bee or a queen bee caste for the rest of their lives. This change creates either a queen bee (female), which can lay thousands of eggs a day for years, or a worker bee, which can pollinate hundreds of thousands of flowers over its six-week summer lifespan. This change also allows a worker bee to perform numerous progressive duties inside the hive and in the local neighborhood. A queen bee cannot change into a worker bee, nor can a worker bee become a queen. These two types of female honey bees form the social caste in the colony. Each member of the honey bee caste has specialized social duties that only it can perform. These permanent social roles are called **social castes**.

Drones are male bees; they are physically larger than female bees and unable to perform the duties of either type of female bee. The drone is not included in the caste and performs no duties like either type of female bee. Drones are fatherless and independent of male parents. These male bees are called **haploid** because they only have 16 chromosomes from their mother. Female bees are **diploid**, as they have both parents and 32 chromosomes.

The Honey Bee Colony

Life Begins in the Brood Nest

The pre-adult bees have their beginnings as eggs in the brood nest. After three days as an egg, a rice-grain-sized larva hatches from the egg. The birth cell is capped with porous wax for incubation, where the larva progresses to a pupa. This pupa is in capped cells for about 10 days before using its newly formed mandibles, or jaws, to break the cell capping to merge. This rite of passage for a bee is the point at which it becomes an adult bee. Newly emerged adult bees look blond and smaller than an older adult foraging bee. After a couple of days, this young bee will appear the same as other worker bees in the colony.

All female eggs start off the same. After three days, these eggs hatch into larvae. All female larvae are still equal. On day three, nursery bees change the diet for the larvae. All queen larvae continued to be fed the protein-rich royal jelly, but worker bees were weaned from royal jelly and fed nectar and honey. This change in diet creates a worker bee. Drones follow the same diet path as worker bees, but a drone egg can only become a drone. Both a worker bee and a queen bee can create drone eggs. Worker bees are unmated but can create unfertilized male eggs; these worker bees are called **laying workers**. In a healthy colony, these laying worker eggs are cannibalized by worker bees, so only the queen's offspring are cared for.

Queen Bee

*A single queen mother exists in each healthy colony.
She is surrounded by her offspring or adopted family.*

A mother queen can live up to five years. Queens can die for multiple reasons; can be an intentional death, while others can be accidental. She can die naturally due to less-than-ideal genetics, or she can be killed by her colony, succumb to disease, beekeeper error, fall to hive intruders, fatality during mating flights, chemical exposure, not being fed or starved of protein, overheating, not enough attendants to care for her, winter fatality, and perhaps more could be added to this list.

The queen is the only bee in the hive with mature, fertilized ovaries. She is the only bee in the colony capable of creating female worker bees. She has a sac-like organ the size of a pinhead called a **spermatheca**. This sac stores sperm from 10–20 drones for her entire life. She collects the drone sperm during consecutive mating flights in an area called a drone congregation area (DCA). Queens initially leave the hive for mating, but once she returns, she stays in the hive for the rest of her life, except for swarming and absconding.

Queen bees can be artificially inseminated in a lab with drones' semen, creating a fully mated queen with ideal genetics. Her incubating cell is the largest of all the cells in the hive and is the only cell that opens downward. The queen is the largest bee in the colony, with an abdomen twice the length of a worker bee.

All female eggs that a queen produces have a chance to become either a queen bee or a worker bee. These female eggs stay intact for three days, then larvae emerge. This larva is fed a whitish substance called royal jelly or bee milk. The tiny larva eats and swims in the tiny pool of royal jelly for four days. After day four, only the queen bee larvae are fed this royal food. All other larvae that are weaned off the royal jelly diet will become female worker bees.

When the queen is laying eggs, her attendants care for all of her needs. A healthy queen can lay up to 2000 eggs a day during the foraging season. She is groomed, fed, protected, and even has an excrement crew removing her poo. She does not defend, clean, or search for and collect food. The queen's only task is to lay offspring.

Being a female, the queen has a stinger, which is called an **ovipositor**. Through evolution, the honey bee stinger no longer has any function related to reproductive use. Other insects use their ovipositor to inject eggs that later hatch and feed on the decaying carcass. The queen bee also releases a pheromone called a queen's substance or her queen's essence. She releases this scent through her mandibular glands, shared by the young nurse bees. This pheromone is the social glue that holds the colony together. Each bee can smell their queen's essence, or her scent. If the queen's essence fades or is lost, the queen is dead or gone. There are situations where a colony can lose their queen. Ways a queen dies are as follows:

- natural death due to less-than-ideal genetics,
- killed by her colony as something was not ideal to them,
- can succumb to disease,
- beekeeper error can accidently cause her premature death,
- hive intruders can kill her,
- she can die during mating flights,
- exposure to toxic chemicals near the hive, or in the pollen she eats,
- not being fed due to too few bees to collect resources, or is starved of vital protein,
- she can overheat if the hive it too hot, and bees are unable to bring in water for cooling,
- there are too few attendants to care for,
- can die in winter due to cold, lack of food, disease, and perhaps more could be added to this list.

Worker Bee

Versatile worker bees live short lives yet perform a vast array of hive duties.

In a beehive, there are three types of bees: one is the queen bee, another is the drone, and the third is the worker bee. What does each of these types of bees do? Each bee type is essential to the survival of the colony, but the worker bee appears to be a multi-tasker. She can go from job to job effortlessly throughout her life. From the moment a female worker bee leaves her birth cell, she works. She evolves and progresses through progressive age-based tasks effortlessly and continues for her entire life. The change between the female queen bee and the female worker bee occurs not in the egg but four days later, when it hatches into a tiny white larva. This rice-grain-sized egg sits in its very own cell in an area of the hive called a **brood nest**, also called a **brood chamber**. Up to day three, all larvae are fed a bee-secreted protein-rich superfood called "**bee milk**" or "**royal jelly**." After the decision of the nursery staff in the brood nest, worker bees are chosen and fed a honey diet. This change in diet creates a smaller female worker bee in 24 days. The larger queen sisters were not weaned off their royal diet and, in 16 days, grew to be larger bees with the ability to mate with drones. This amazing instant change in diet paves the way for this working bee to serve the colony the rest of its life in almost every capacity a bee performs. A worker bee can live five to six weeks during the foraging season. During the plant winter season, a worker bee can live for five months or more.

Drone Bee

Drones are mating specialists. The drone is a bee with only one task, but a crucial one: mate with available nearby queens. The only male bees in a beehive are drones. They are not equipped with stingers, honey guts, or pollen baskets. They are visually larger than worker bees, and their abdomen ends square instead of with a point, like females. A drone has larger eyes than a worker bee, and their antennae sit lower on their heads. Each healthy colony has ideally 200 or more drones, who are created beginning in spring and are permitted free access to the colony's protected honey reserves until fall. Their job is simple: to stay healthy, provide body heat for the colony when clustering during the season, offer competition to other local drones, and mate with a virgin queen when available.

During the midday hours, drones head to their fraternity hangout called the drone congregation area (DCA). This area, high in the air, is where queens fly to be mated. Drones live from spring to fall. They die when they successfully mate or are expelled from the hive in late fall to save resources for the winter worker bees. These evicted bees die from starvation, hypothermia, or being prey to local insect-consuming animals. When a drone is not feasting on honey in a hive, they can fly 3–5 miles (5–8 km) in any direction in search of other honey-stocked hives to eat from and other DCA areas to mate in. They inadvertently migrate with their queen mother's DNA briefcase to distant colonies. This way, they share unrelated genes with the new queens. Drones are welcomed as visitors by all colonies prior to the natural fall eviction deadline when they die.

Drone honey bees also contribute to the thermoregulation of the hive, or internal environment. They use their bodies to help support an ideal internal temperature in the hive, which is necessary for the survival of the colony. Another unique feature of a drone is that it has no father, only a mother. When a queen creates a drone, the sperm she uses is not fertilized with drone's sperm, so it is called **haploid**, or has only one set of chromosomes. Female bees are diploid as they have two sets of chromosomes, both contributed by a mother and a father. Each spring, about three weeks prior to the swarming season, the queen mother begins by laying patches of drone eggs.

Mating bee specialist. The male drone has one duty: to mate with a virgin queen.

The World of the Queen Bee

Queen Bee

The queen emits a cohesive pheromone called the Queen's Essence. The bees are all scented by her, and this scent shows their home.

A mated queen is the only bee in the hive that can lay eggs that become female bees, the future queens, and worker bees. This queen can also choose to lay male eggs; they become drones. Even though the queen has an ovipositor as a stinger, this reproductive organ, through evolution, is no longer used by honey bees for laying eggs. The ovipositor in honey bees is solely a venom-packed stinger. Female wasps still use their ovipositor to lay eggs on whatever they sting. The queen stays in the brood nest, laying eggs, and avoids venturing out beyond the honey. Nurse bees and attendants to the queen remain in the brood nest. Each queen is equipped with an interior sac called a **spermatheca**, which stores semen from all the drones the queen mated with to be used throughout her life. The queen's pheromone is also known as the queen's substance. This substance is what bees use to show their queen and colony. This substance is secreted by the mandibular glands of the queen and is the social glue for her colony.

Adopted Bee

Adoption requirements exist for unfamiliar honey bees to be accepted by a new colony family. These requirements apply to the queen, worker, and drone.

Adoption agencies exist for bees, supported both by beekeepers and bees. Queen bees are regularly placed in an adoption setting when inserted into a small queen cage with several of her caring offspring attendants and combined with thousands of unrelated worker bees for four days. Initially, the foreign, adopted queen is seen as a threat due to her unfamiliar pheromones. After four days, the original queen's pheromone has faded and is replaced by the newly adopted queen's pheromones. Once the bees learn the new scent, they accept the new queen as their own mother queen.

Worker bees, too, can be adopted into a new or otherwise hostile colony. Worker bees can get the scent of the new colony just by hanging out on the exterior of the hive. A fast-track avenue for adoption into a foreign colony is when the new bee comes bearing gifts, such as resources like water, pollen, propolis ingredients, or nectar.

Drones are universally accepted by all colonies unless the colony is preparing for winter. At winter's doorstep, all drones are forced out of the female-controlled hives and treated as outcasts. These drones die of the cold, starvation, or by being eaten by visiting predators.

Fighting Virgin Queen Sisters

Young virgin queens often fight to decide the next queen mother of the colony.

When colonies requeen themselves, multiple queens can be produced. These queens are called **queen sisters**. When multiple queen sisters are raised in a hive, they can fight for sole dominance of the colony. This royal battle is announced by the queens with an audible sound called **piping**, **tooting**, or **bugling**. It may sound like a chirp from the outside of the hive. After the battle, only one queen will remain in the hive and be accepted by the colony. The losing queen will be ousted and beaten by worker bees near the hive.

Royal Executioner

Efficient supersedure is when the queen is failing and a new daughter queen is laying eggs, so one queen must be eliminated. Queen balling usually occurs when an old queen is being replaced or a new queen is improperly introduced to the colony. This agitated group of hissing worker bees is clinging to the queen like a scrum in rugby. In the center of the ball, they are stinging, suffocating, and overheating the queen instead of protecting her. Bees breathe through air sacs on the abdomen, so the balling bees are cutting off her air supply. Closest allies to the queen unitedly turned on her like the senators who killed Caesar on the steps of the Senate in Rome.

Bees can turn on their queen when she is injured, unhealthy, or unable to lay eggs.

Royal Excrement Bee

Young bees constantly groom and clean the queen, including her poo.

Bees confined to the hive hold off pooping for as long as they can. Like a dog in its kennel, it will hold off until it can go outside. For bees, they can hold their poo for an exceptionally long time. The queen's job is to lay eggs constantly. She is fed by those attending to her needs. She also has bees who groom her and keep her clean. Cleaning involves licking up her poop. Again, as with dogs and other animals, excrement is not disgusting to bees. Cleaning poo is a way to share respect and care for another. Fish swim in their excrement, while some insects, like flies and the dung beetle, consume it. So, hold off on the development of a diaper for a queen bee. The bees have it covered.

Swarming Bee

Swarming with their queen to a new location is a rare and exciting event for a bee.

Swarming is planned long before the bees leave the hive. Preparation includes thinning the current queen, as she will be leaving with the swarm. This old queen needs to be light enough to fly. Also, the old queen needs to lay a new queen. Usually, queen cells are left behind in their capped pupal stage. When the queen cells are in the middle of the brood nest, they are referred to as a **supersedure cell**. This type of queen was created due to an old queen who was failing, injured, or killed. If the queen cells are found at the bottom edge of the brood nest, they are called **swarm cells**. Nurse bees choose these queen cells because they are the only place left for larvae that could be made into queens. There was no more room left to expand. Once a swarm cell is capped and incubating, the old queen leaves the hive for the young new queens to take it from there. These young queens can fight it out, or the first queen to emerge may visit her incubating sister queens and kill them while they are incubating inside a capped cell. These queen battles create an alpha female queen.

Trojan Horse Bee

Africanized honey bees use old world war tactics in overtaking an unsuspecting colony.

Africanized bees swarm with a young virgin queen and seek out existing colonies to conquer. When these angry, rogue bees find an occupied, established colony, they stay hidden near the hive entrance until they have the colony's pheromones and are seen as family. Once they are granted access to the hive, they seek out the queen mother and kill her. When the unsuspecting colony is motherless, the intruders escort their queen in to be adopted by the new colony. Africanized bees can be identified by their DNA, excessive aggression, and the cell pattern in their wings.

Vibrating the Queen Bee

Prior to swarming, the queen's attendants poke and vibrate her to lose weight.

Even a queen bee must exercise. When a colony has reached its population limit or has decided to abandon its home, the queen needs to exercise. This exercise helps prepare her to lose weight prior to the trip. Attendants to the queen help the queen by vibrating or buzzing her to keep her moving. This continued movement allows the queen to lose weight prior to using her wings for flight. The heavier the queen, the harder it is for her to take flight and sustain herself over a long distance. Scout bees guide the swarm to a new location. The queen needs to be healthy, and strong prior to this strenuous relocation. This explains why swarms can be found on the ground or on low branches, compared to other swarms that can reach the top of a 60-foot (20-meter) tree.

The Amazing Worker Bee: Activity in the Brood Nest Nursery

Nurse Bee

Nurse bees are young bees who clean baby cells, produce wax and royal jelly, care for all the needs of the queen, and control whether the queen is to remain or be replaced.

If there was a ruling class in the hive, it would be the adult nurse bees. Nurse bees' control and care for the brood nest, the inner sanctum of the hive where babies are made, and the queen resides. Nurse bees are 3–10-day-old adult bees. They clean their birth cells, feed larvae, and help regulate the humidity in the brood nest. These young bees oversee all the needs of their queen mother. Nurse bees can function as heater bees, keeping capped pupae warm by entering nearby empty cells and generating heat. All older bees serve the needs of the brood nest, while the nurse bees intimately care and maintain it. The brood nest is the most protected and regulated area of the hive.

Nurse bees will inspect growing larvae thousands of times in a single day. One reason honey bees inspect brood cells is to ensure that the developing larvae are healthy and disease-free. Worker bees can detect signs of disease or infection in developing larvae and will remove them from the colony to prevent the spread of disease. By regularly inspecting brood cells, bees can catch diseases early and prevent them from spreading to other members of the colony. Honey bees inspect brood cells to ensure that the developing larvae have enough food to survive. Bees feed developing larvae with royal jelly, nectar, and pollen, depending on the age and need they have for the larvae. When the nurse bees detect a failure in the queen or if she is gone completely, they will decide which young larvae is to become a future queen candidate. These same caring nurse bees also decide if their queen is to be replaced.

Attendant to Queen

They are cared for by the queen's royal court, which consists of young nurse bees. They care for their queen mother.

The **royal court** is a small group of 10–20 worker bees who surround the queen mother. This small group is called the **court,** or **attendants,** or the **queen's retinue.** The queen's youngest offspring care for all her needs. They groom, feed, and perform waste removal. These nurse bees are the first to be exposed to the queen's pheromone, called the **queen substance**. They care for the queen and can range in age from a few hours old to fully mature bees. Her main diet of royal jelly is supplied continuously by her attendants. In the bee world, the mother does not do the cleaning in the hive. She does inspect each cell prior to laying an egg in the bottom center of the cell. She does not undertake worker bee chores. The queen's only job is to make more offspring.

Bee Bread Pounder Bee

Bees convert and pound fresh pollen into the colony's only protein, called bee bread.

In a beehive's nursery, called the **brood nes**t, there are eggs and larvae in disinfected wax cells. These brood cells are surrounded by cells stocked with vital nursery supplies. These supplies are bee bread (fermented pollen as protein) and honey reserves (as carbohydrates), propolis (disinfectant), and water for adjusting humidity.

Foraging bees collect fresh pollen and nectar from flowers and return it to the hive. Receiver bees take them, add enzymes to them, and store them in cells near the brood nest. The pollen is also pounded, fermented, and packed with their heads into cells. Bee bread is essential to producing royal jelly. As nurse bees eat the colored bee bread near the brood nest, they secrete royal jelly from their hypopharynx glands. This royal jelly is then used to feed young larvae as well as their queen. A queen lays up to 2000 eggs a day, and this requires lots of continuous royal jelly protein in her diet.

Brood Heater Bee

Nurse bees enter empty cells in the brood nest, where they vibrate to induce warmth for the surrounding cells.

Heater bees maintain the temperature of the brood nest. Empty cells in the brood nest may appear to be an unproductive use of valuable hive space. Nurse bees use these empty cells to enter to help maintain a brood temperature of 91°F–96°F (32.8°C–35.6°C). Young bees are sealed into wax cells while they develop. Brood nest heater bees insert themselves into an unused nearby cell. While in the cell, the bee vibrates to stimulate heat in the surrounding area. Honey bees regulate hive temperature by generating heat; this is called **thermoregulation**.

Cell Capper

Worker bees cover larvae in preparation for incubation and ripe honey with wax for future use.

Capping cells is done for two reasons. One reason is to seal ripened honey, and the other is in preparation for larvae progressing to the stage where they are incubating. Once ripened nectar is dehydrated enough to become honey, bees cap the cell. The ripened honey has less than 16% moisture. This is the bee's way of preserving their food for times of famine and winter. Cooked honey wants to rehydrate or absorb moisture from the surrounding environment. This is a hygroscopic quality of honey. Capping honey with wax also prevents any contamination of the stored preserves. When there is a need for food, bees simply tear open and recycle the cell capping. Capping cells with larvae inside allow the incubating pre-adult bees essential private time alone. Worker bees cap the cells of nursing larvae prior to progressing to the pupa stage. Once the fully formed bee is ready to emerge, it gnaws its way through the capping to become an adult bee. Essentially, this uncapping is a pre-adult bee's rite of passage to become an adult bee.

Childcare Nursery Inspector

Nursery inspectors care for eggs and young larvae.

In a beehive's nursery, called a brood nest, there are eggs and larvae in individual cells. This nursery is surrounded by cells stocked with vital nursery supplies, bee bread and honey. The bee bread is their protein, and the honey is their carbohydrate. Royal Jelly is not a stored resource but is continuously made fresh by young bees.

Childcare workers can visit an occupied cell 1300 times a day, and up to 10,000 times during the eight-day period prior to the cell being capped. During these visits, bees can make a brief visit, or make a full inspection, or they can feed the young. Visits last two to three seconds. Inspections are about ten to twenty seconds. Feeding can take up to three minutes. https://www.beeculture.com/a-closer-look-nursing-behavior/

Cell Size Bee Inspector

Front-facing cells can be used to store resources or as incubation sites for worker bees and drones. Down-facing cells are only for future queen bees.

Initially, when bees are building wax cells in a hive, they appear hexagonal with six sides, but by the time the cell is finished, the openings are circular. These cells can be used for incubation or food storage. A brood cell for incubating a worker bee can vary from 4.62 mm to 5.51 mm in diameter. A drone nursery cell can vary from 6.15 mm to 6.91 mm across. Queen cell diameter is about 8 mm to 9 mm. The upward incline of the cell is also necessary, as it helps keep the liquid contents in place during processing and storage. This angle varies upward from 0.05 degrees to 19.25 degrees, with an average upward angle of 7.3 degrees. Queen cells are made on demand and differ completely from all the other cells in size. Queen cells are the largest cells in a hive. All wax cells in a hive open outward except for queen cells that face downward. Laying worker bees lay their eggs in worker-sized bee cells. Drones raised in worker cells are smaller due to the size of the nursery cell.

Laying Worker Bee

The difference between a queen bee and a worker bee is that a queen has developed ovaries capable of producing female eggs. A worker bee has underdeveloped ovaries incapable of creating fertilized eggs. All unfertilized worker bee eggs become drones. A drone is a male bee that does not have ovaries and is incapable of egg production. A mated queen also emits a mated queen pheromone, known as the queen's essence. No matter how much a worker bee wants to replace the queen or make up for a missing queen, she does not have the reproductive hardware to do so.

When a queen goes missing and is gone for more than 10 days, workers begin laying unfertilized eggs. These worker bees are called **laying workers** and identify as queens. Each colony can have multiple laying worker bees at one time, who are confused about who they are and their personal reproductive limitations. They essentially think they can reproduce, but they cannot. Compared to a mated queen, a laying worker lays multiple eggs in a birth cell, which are poorly placed. A queen bee usually lays only one egg per birth cell, which is placed perfectly in the bottom center of the cell.

When unfertilized eggs are produced, this throws the natural balance of the colony into chaos. The drone population skyrockets, creating a population of drones who are unable to defend, produce, or work. All worker bee duties begin to stop. Drones are only focused on consuming honey reserves. Soon this colony implodes as those producing all the resources age and die with no replacements. Honey reserves are eaten by the increased number of fatherless bees, incapable of performing any sustainable labor. Soon, only drones will remain. These drones dissipate by visiting other nearby colonies, which allow them temporary entry, or they die. This situation is called a **laying worker condition**. When a queen is added to this confusion, her life can be threatened by the workers who think they are queens. Laying workers do not emit queen pheromones, so the social glue holding the colony together is lacking.

Laying worker bees who are confused about their self-identity and pretend to be queens. This is a sign of a failing, dying colony.

Propolis Cell Polisher

When a bee emerges from its birth cell, it cleans and disinfects the cell with propolis.

Bees do not make beds and tuck linen sheets, but they do clean and disinfect the nursery cell they were incubated in. Each bee raised in the cell polishes it after use with propolis. This polishing darkens the wax walls with time. The propolis polish is antibacterial and antifungal. With time, the brood nest cells appear darker and darker after repeatedly raising young. This propolis polish is a beehive disinfectant, so the next egg is placed in a sterile, clean setting. Before the queen lays an egg in the cell, she inserts her head and smells it with her antennae to ensure it has been prepared.

Royal Jelly Maker

Royal jelly is a protein-rich bee milk for larvae and the queen.

Royal jelly, or bee milk, is made by young bees in the hypopharyngeal gland. This gland is behind the bee's head. Royal jelly makers produce and share this essential, translucent white jelly-like protein. It is rich in nutrients and is fed directly to bee larvae. After three days as larva, only queen larvae continue to receive royal jelly. Royal jelly is not a stored resource but is continuously produced, is high in protein, and is generated when worker bees consume stored bee bread. Three nursery essentials: bee bread, honey, and royal jelly.

Wax Chewer

Glands in the bee's underbelly make tiny wax flakes, which they chew to soften and use throughout the hive.

Wax is produced by young bees through their wax glands located underneath their abdomen. Wax production is associated with high nectar collection and is followed by an increase in eggs laid by the queen. As bees age, their wax glands atrophy and stop producing wax. In preparation for swarming, bees begin producing wax for their new home. Once wax is secreted, other bees take it like fruit being taken off a tree. They chew the new wax flake with their mandibles and take it to where they think is the best location. Another bee may, in turn, chew it again and move it to another location until it is finally left as a part of a cell, a capping, or other supporting hive feature. This is the process of how honeycomb wax is formed.

    Wax is a renewable resource in a hive. During times of low wax production, worker bees may recycle and repurpose wax. The wax pheromone stimulates nesting behavior. Beeswax is composed of esters of fatty acids and long-chain alcohols. It can be used in candles and burns cleaner than petroleum-based candles. Beeswax candles have been used throughout history for heat, light, seals, balms, makeup, hair products, and coatings for wood and leather.

Worker Bees Jobs Inside the Hive but Outside of the Brood Nest

House Bee

House bees are all the bees inside the hive performing jobs beyond the brood nest nursery. All the jobs of a worker bee are in support of the brood nest. All bee duties are age-progressive and are signaled by pheromones. As bees age, their progressive duties take them further away from the queen and her nest. The most secure area within a honey bee colony is the brood nest.

House bees are those with duties beyond the nursery.

Biomonitoring Bee

Bees help researchers, who evaluate the amount of chemicals that leach into a bee's diet by studying the pollen or nectar that is brought back to the colony. The bees, and harvested resources in the hive, are inspected prior to chemical exposure, then again after exposure to figure out how the chemicals affected the bee's health.

Scientists have united with honey bees in studying the environment. Bees can provide valuable environmental insights to researchers, like canaries provided safety in a mineshaft. Mining companies used canaries to alert unacceptable carbon monoxide levels, which is invisible and deadly to both the bird and the miners. Honey bees rely on a 2–3-mile diameter from their hives for life sustaining nourishment. Like the canaries in a mine, honey bees are invaluable in determining the safe use of products used in the production and manufacture of food and products humans need and desire in their lives. Bee pollen (protein), nectar (carbohydrates) and wax (animal fats) are routinely tested to ensure bee health is maintained, as well as a suitable end-product is produced for consumers. When crop chemicals are initially produced, field tests look for results that help one of the participants and hurt neither. Or where both participants benefit. This is called **mutualism** (helps both) and **commensalism** (helps one but does not harm the other). Biomonitoring is a way to see that bees are healthy, and their surrounding areas are safe for them to live in.

Blower Bee

Bees do not vacuum their hive, instead they use their wings to blow and push debris out of the hive.

Bees use their powerful delicate wings as they circulate air throughout the hive. This powerful wing air circulation system is used to move debris, rather than picking up each individual particle with their mandibles. After an extended period isolated inside the hive, clumps of blown waste can be seen outside of the entrance. Blowing is an especially useful exercise to strengthen the thorax muscles prior to becoming a foraging bee. The thorax is the center portion of a bee's body and is considered the engine of the bee.

Congress Bee

A bee colony is a superorganism; every job is essential to its success. There are no elections within the hive, nor self-oriented political officers. Bees act in the interest of the colony.

There is no ruling class in a beehive. If there is any word to define a colony of bees, it would be a communal society. In a communal society, the focus is on the group rather than the individual, and all things are jointly shared. This is like the Amish and other religious social orders, where all things are held in common. Bees serve for the benefit of the colony, not for self. All bees have a united voice, colony pheromone, and distinct vibration. It may appear that the queen is the ruler of the colony, but she can be replaced by the very nurse bees caring for her. Perhaps the worker bee is the ruling class, but they go from job to job effortlessly, ultimately dying in defense of the hive or alone. A drone holds no control other than waiting to mate with a future queen.

Cook Chef

After sipping nectar from a flower, bees return the collected nectar to the hive, where enzymes and pollen are added prior to cooking it.

Bees begin to convert nectar into honey by evaporation while it is still on the tongue. This process reduces moisture in the nectar. The elimination of excess water from nectar requires considerable energy. Another way they dehydrate nectar is by fanning their wings, which creates warm airflow around the honeycomb, which helps moisture evaporate from nectar. Once the nectar has ripened into honey, it has so little water that no microbes can grow in it. The hydroscopic nature of honey looks to rehydrate into nectar, even by absorbing moisture from the air. So, after it is bee-cooked, it is capped with wax and stored for times of drought and winter.

Disaster Cleanup

Bees will not tolerate comb damage. Leaking, open cells are at once cleaned and repaired.

Bees have a natural drive to keep order and cleanliness in their hives, so if they detect damage to their comb, they at once begin the repairs. Minor repairs are rectified quickly. But when the damage is severe, the bees may require resources to repair the damaged combs. Bees need nectar to produce wax, which they use to construct their combs. If the hive is low on resources, or nectar is lacking at that time of year, the bees may not have enough wax to rebuild the combs.

    The age of the bees also affects their ability to repair combs. Younger bees have more active wax glands and are better able to produce wax to reconstruct combs. Wax glands in older bees may have atrophied, making it more difficult for them to repair combs. Severe comb damage when resources are low may cause the bees to completely abscond. Any honey leaking from the hive attracts robbers, which requires defense. Too many robbers may induce the colony to move to a safer location.

Disease Inspector Bee

The nursery is where disease may be found. Bees perform health inspections by tasting and smelling the infected bee. Once disease is found, the area is cleaned and disinfected.

Bees investigate death in capped cells emitting kairomones, a scent that stimulates a defensive action. Bees will perforate capped cells to investigate. At times, there are issues with dead larvae or pupae. When the birth cell is capped, bees will bite a small hole in the wax cap and insert their tongues to sample the liquid inside. The color, viscosity, and odor of the liquid can supply information to the bees about the health and nutritional status of the pupae. This allows the bees to make decisions about whether to continue caring for the sealed brood or to remove and discard it. Once the pupa is dead, it decomposes inside the cell.

Dock Worker

When a bee returns to the hive with resources, they hand it off to specialist receiver bees, who send it to where it is needed.

Worker Bees Jobs Inside the Hive but Outside of the Brood Nest

You will not see forklifts or semi-trucks pulling up to the front of the hive, but it is a highly active warehouse dock with resources returning constantly during warm daylight hours. The bees returning are not the ones to take the products and cook, store, ferment, and ventilate as needed to produce the final products. Receiver bees are the go-between to ensure resources end up where they are needed.

Engineer Bee

Bees hang onto each other, creating a bee scaffold for others to walk on while a comb is constructed.

Communication holes built into the comb allow bees to move from one comb side to another. The engineering inside a beehive is a complex and well-coordinated system that has evolved over millions of years. The hive is essentially made up of a series of interlocking compartments, each with a specific function and purpose. The hive is constructed out of wax that is produced by worker bees. The wax is formed into hexagonal cells, which are used to store honey, pollen, and brood. The hexagonal shape is incredibly efficient, as it allows for the largest amount of storage space while still supporting structural integrity.

The most impressive aspect of the engineering inside a beehive is the way in which the bees work together to support the overall health and well-being of the colony. Bees will form scaffolding with their own bodies, clasping one leg to the leg of another bee. This allows bees to scurry across them to perform their jobs. The act of bees holding together is called **festooning**, also called **chaining**. Open spaces in construction zones are often filled with bees, who essentially function as bridges to the front lines.

Fanning Bee

Fanning is an essential activity for honey bee colonies, and it serves essential functions, including thermoregulation, ventilation, and communication. When bees use their bodies to cool or heat the area around them, this is called **thermoregulation**. The primary purpose of fanning is to regulate the temperature and humidity inside the hive. As bees work together to fan their wings, they create a steady stream of air that circulates through the hive, removing excess heat and humidity. In this way, fanning helps support best conditions for brood development and prevents the buildup of harmful gases that could threaten the colony's health. To maximize the effectiveness of fanning, bees often station themselves at strategic locations throughout the hive, such as near the entrance or close to the honey stores.

Another crucial function of fanning is ventilation. Bees use their wings to circulate air through the hive, allowing fresh air to enter and stale air to exit. This process helps to remove excess carbon dioxide and other gases produced by the bees and promotes a healthy oxygen balance within the colony. Fanning also serves as a vital form of communication for the bees. When bees fan their wings, they share pheromones that convey information to other members of the colony. These pheromones can signal the queen's presence, danger, the availability of food, or other vital information that helps the colony function efficiently.

Worker bees share their queen's pheromones by fanning their wings.

Grooming Bee

Hygienic grooming behavior can be seen when a bee removes a threat from another bee.

A grooming bee school is needed for honey bees. Grooming behavior—where a bee cleans another bee—is not standard behavior. For the most part, bees let other bees take care of themselves, except for the queen bee and returning foragers. A worker bee performs a grooming shake dance to stimulate another receiver bee to remove dust, sugar powder, pollen, or mites from the dancing bee's body. Russian honey bees, groom each other and are more resistant to the threat of varroa mites. Hygienic bees are believed to be the answer to defending against the mite. Researchers are currently working on raising bees in labs that are resistant to mite threats. Russian honey bees and Africanized honey bees appear to be naturally more hygienic than others, but the Minnesota Hygienic Bee has been showing an ability to defend and eradicate varroa mite populations in their hives.

H-VAC Bee

The bees manually do H-VAC or heating, ventilation, and cooling.

The environment inside the hive is manually kept by bees. In winter, the bees shiver and shake to generate heat as they cluster together. Bees do not hibernate. They are lethargic but still active all winter. As they cluster in winter, bees slowly rotate from cold locations to warmer locations in the cluster. Bees eat capped and unripe honey during the winter, but unripe honey may have higher moisture. When bees consume unripe honey during the winter, moisture is produced. This humidity can freeze and kill bees during their winter isolation.

In spring and summer, bees spend time cooling and heating the hive, depending on external conditions. The humidity is most concerning for the eggs and uncapped larvae. When the humidity drops below 70%, eggs and larvae mostly fail. The optimal humidity in the hive for the young and eggs is about 95%. Bees collect water and fan it in the hive to produce moisture. This moisture in the air inside the hive keeps the colony cool and the brood nest moist.

Line Dancer Washboarding

This can be found on the front of the hive, with bees in a stationary stance moving side-to-side in rhythmic unison.

Washboarding is a mysterious group activity. No one knows for sure why it is happening. It has the appearance of bees using washboards and scrubbing. Many bees usually participate in rows with their abdomens in the air, their heads lowered, pointing away from the entrance, and their mandibles moving. They each look like a bull, ready to attack. This activity is usually seen later in the day, at the end of a harvest. Many possible ideas for why the bees are doing this exist. Several possibilities include:

- They are showing their readiness for news from a returning shaker bee on available resources.
- This is a unified defense pose warding off robbers, or predators.
- Overcrowding in the hive.
- It is some kind of reverse ventilation.
- They are cleaning "sweeping the porch."
- Bees with nothing to do are hanging out on the porch.
- Bees are marking or polishing the entrance.
- The bees are depositing olfactory cues on the hive.
- It is genetic, as some colonies do it more than others.

Listening Bee

Bees do not have ears but hear with their entire body.

Verbally telling a bee to get out of here will mostly go unnoticed. Bees can see and smell a person's exhaling breath. The smell of your breath may attract defense bees. But certainly, exhaled carbon dioxide can set off a bee into alarm mode. It is best to leave yelling alone and run instead. Bees do not hear with their ears, but they notice the sound with their whole body, especially with their antennae and sensitive body hairs. The bees, during their waggle dance, produce the sound of 250 oscillations per second (250 Hz). Young virgin queens can chirp and make other audible sounds preparatory to a queen death fight.

Master Environmentalist and Herbalist Bee

Flowers are healthier after bees pollinate them. Bees use plant products in the hive for health and to sustain life.

A large bee colony can have 20,000 to 80,000 bees, with a rotation of new bees rising through the ranks every six weeks. With so many bees tightly intermingling socially, health issues can be transmitted very quickly. Honey bees have developed advanced safety measures and hive health protocols. Honey bees are expert environmentalists. As they help nature by pollinating plants, this then improves the health of those plants.

Honey bees are also expert herbalists. Bees use resins from plants for their own medicinal benefit. They innately know how to use the secreted resins of plants to improve their colony's health. These harvested resins form vital ingredients in bee-produced propolis. This propolis is universally used throughout the colony. Propolis can be polished, hard, or sticky. Polished propolis is used in nursery cells in preparation for the cell's next use. Hard propolis can be found on edges. Sticky propolis is used to fill holes and seal edges. This sticky propolis can also function as a defense against intruders. All bees will walk across and track this healing balm throughout the hive.

Mobile Food Services

Bees use their tongues to feed each other.

Worker Bees Jobs Inside the Hive but Outside of the Brood Nest

The tongue of a bee is called the **proboscis**. It is a vital tool for feeding and communication. It is about a quarter inch long, or 6.5 mm. The length of a bee is about two thirds of an inch long, or 15 mm. Comparing the length of the tongue to the overall length of a bee, the tongue can extend to about half the bee's length. The tongue of a bee is an amazing tool. It can lick or suck like a straw. When bees appear to be licking each other's tongues, they are sharing food. This act of sharing food is called **trophallaxis**. Sharing food is also a way bees communicate with each other.

Food transmission exists among all bees in a hive and among all bees from the same colony. Foragers return to feed and share food with house bees. These foragers and house bees also feed hungry drones and their queen, the same way. While young emerging bees are leaving their incubation cells, they hold out their proboscises to be fed. This head-to-head and antennae-to-antennae interaction is how bees share food throughout the colony.

Propolis Applicator Bee

Propolis is used to fill cracks and holes in the hive as a safety and health procedure.

Bees make their own sealant and glue. **Propolis** is a natural resin compound that honey bees produce from various plants. Propolis is a mix of plant resins, bee enzymes, and beeswax. Bees store propolis in their hives and use it to seal unwanted holes. This reddish-brown sticky goo also helps to prevent bacterial, viral, and fungal infections and acts as a barrier against intruders. Propolis is used for health benefits but also has a defensive quality.

According to the USDA Forest Service (https://www.fs.usda.gov/), plant resins are not water soluble, harden when exposed to air, and are generally produced by woody plants. However, resins can be produced by flowering herbs, and buds from a shrub. Often plants will produce resins to form a band-aid over damaged areas. The plants that produce the ingredients for propolis vary but commonly include poplar, birch, alder, and conifer trees. Bees collect the sticky substance from tree buds, bark, and other plant parts.

Recycling Bee

Bees are constantly moving wax that was used in one place to be reused in another location.

Recycling is a common practice for bees. Wax is an essential reusable product in the hive. Wax is secreted from the underbelly of a bee through its wax glands. It is like bee fat. It does not matter to a bee, who made the original wax flake, that it can be chewed and reformed like clay. Wax is the brick and mortar for the colony. When secreted by the bee, the wax is snow white; as it ages and is used and disinfected with propolis, it then turns yellow to deep brown. Once wax falls to the bottom of the hive, it is considered discarded waste.

Relocation Services

Bees have a right and wrong way of storing food. If honey is in the wrong place, they will move it to where they want it.

Bees have a specific way of doing things in the hive. Honey is always above the brood nest, and between the two is a layer of stored bee bread or fermented pounded pollen. When this gets altered, the bees can reshelve honey, but they usually abandon the bee bread. Bee bread is needed to produce bee milk or royal jelly and needs to be next to the brood nest, as that is where it is used. If stored honey is damaged but not needed, bees move it to other undamaged cells they can cap. Capped honey can cause challenges as bees know those cells are finished. If too much capped honey separates brood nests, bees can mistakenly decide their queen has failed and go through the process of creating another queen mother.

Worker Bees Jobs Inside the Hive but Outside of the Brood Nest

Shaker Trembling Bee

When a foraging bee brings resources to the hive, it shakes in areas where unemployed bees are waiting for work. This shake is the bee's way of employing new foraging bees.

When a bee is shaking by itself, it isn't scared; it is performing the tremble dance. This dance is the bee's way of trying to recruit unemployed bees. This bee is communicating an urgent need and is showing nearby relatively motionless bees to pay attention to the new job offering.

Sleeping Bee

Bees sleep for short periods where they lie motionless.

Bees do sleep! When honey bees sleep, their antennae droop motionlessly, their abdomen rests flat on the ground, their wings lay tight against their abdomen, their legs straighten, and their body temperature drops. They can sleep 5–8 hours a day. If they are alone, the bees can even rotate sideways. If sleeping with other bees, they hold each other's legs, which prevents the bee from rolling over. Bees sleep face down, flat, and motionless. Young bees inside the hive have no specific nap time. Older bees who have left the hive during the day return to sleep with their colony at night. When foraging bees lack sleep, they lose the ability to learn new waggle dances and can struggle to find their way home.

Traffic Control

Traffic at the entrance of a beehive may seem frightening to a modern aviator, as bees return and take off quickly.

With poor eyesight, it is amazing that the honey bee can come and go from their hive as fast and effectively as they do. Do bees crash into each other in the air? They certainly do, but it is a rare occurrence. There is not a traffic control tower or safety patrol governing bee flights and landings. A large colony can run with only a finger-sized hole, but larger entrances serve the needs of larger honey-producing colonies better.

Thermal Defense Bee

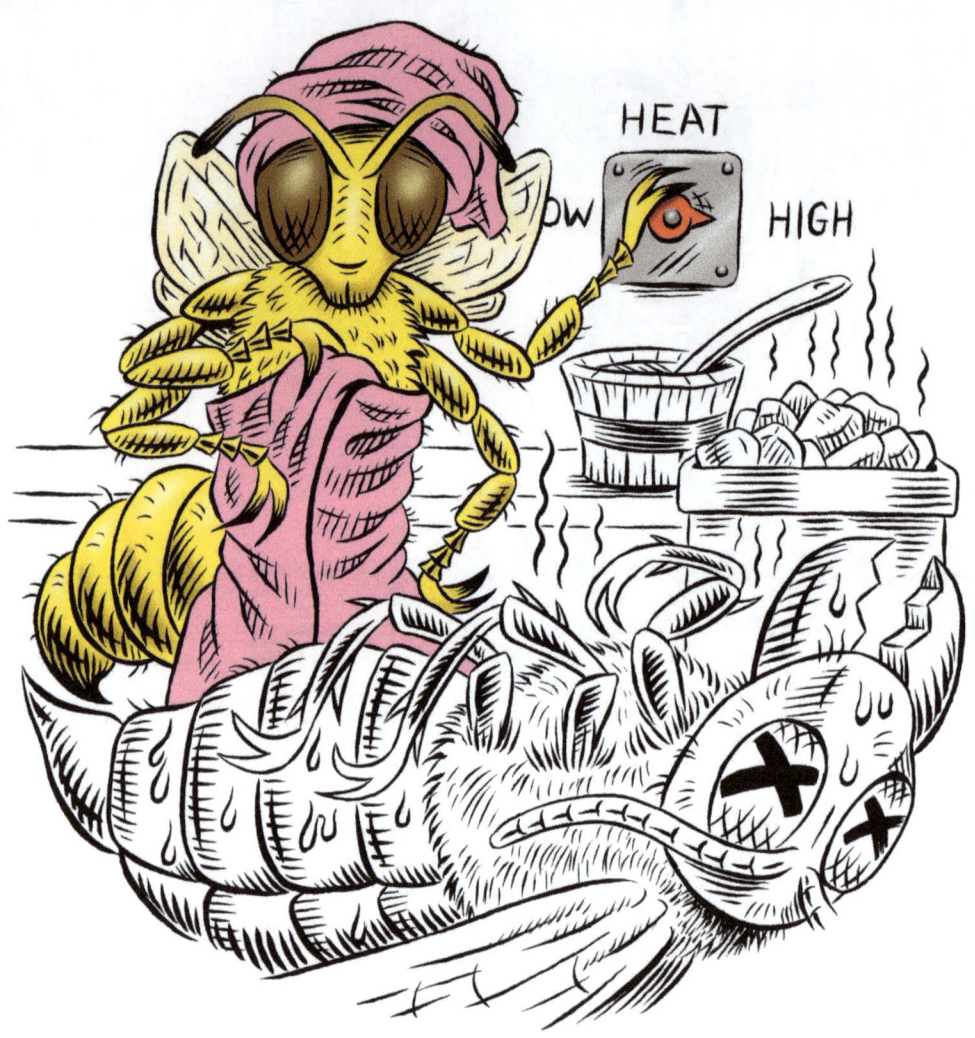

Bees can use heat as a defense against invaders.

Honey bees have a social structure that allows them to work together to defend their colony. A colony's defensive tactic is to surround the intruder in a heat ball. Bees can generate heat by vibrating their wings in a process known as shivering. When bees shiver, they can increase the colony temperature by over 20 degrees, to as much as 114°F (46°C). This united "heat ball" defense works, as hornets then disorient and overheat. Increasing the hive temperature has also worked to defend against Varroa mites. When heat is increased for a few hours to 108°F (42°C), the tiny mites overheat and die. Bees using heat as a defensive weapon is a remarkable adaptation in the bee colony.

Unemployed Bee

Unemployed foraging bees wait in line for advancement.

Do bees wait in lineups for work? Not really lineups, but they will congregate in areas where returning foragers visit. These waiting worker bees have progressed from nurse bees to duties inside the hive and now have advanced in age to be a field bee. It is not as easy as just flying out and doing whatever a bee thinks it wants to do. Other field bees and scouts return with news on where to find resources. Once the bees accept the news and learn directions from the other bees, both in movements and scents, the new field bee leaves for its new journey serving the colony.

Waggle Dancer

The waggle dances are communication between bees. This activity is done by circular movements, pheromones, and shaking.

Body language, eye contact, and vocal patterns can share human messages. Honey bees have two primary methods of communicating with each other: movement and odor. Bees also have body language through which they send messages; these are found in their types of waggle dances. These messages can instruct other bees where a needed resource can be found, as well as supply other information. Dancing is a series of repeated movements of bees on combs and in swarms. Waggle dances include the round dance, the sickle dance, the wagtail dance, and the shake dance. Returning worker bees perform a variety of creative dances that, through action, are a road map. Bees then follow the clues, like finding a bee's geo-cache location. It is difficult for a bee to find resources close to their hive, but they can. Nearby water can be found by bees as close as a few feet away. It takes a bee about twenty feet, or six meters to attain their ideal flying height. After performing a waggle dance, the scout bees may share the foraged food with the following workers to communicate the quality of the food supply available at the location.

Winter Bee

In autumn, the colony's queen mother lays eggs that will become winter bees.
These bees can survive in the hive until spring.

Winter bees are also called fat bees. Winter bees do not fly seeking nectar in winter; they can survive winter inside the hive until spring. During the warm months of the year, bees can only live up to six weeks and are knocking at death's door. These winter bees huddle together at the bottom of the hive on layers of capped honeycomb. While in the winter cluster, they eat ripe honey, which is a carbohydrate with low moisture content, to produce energy and shiver. This shivering is not because they are cold; it is to produce heat. Together, they all shiver, creating a temperature of about 92°F (33°C). As the bees on the outside of the cluster cool, they move back inside. This thermoregulation is vital in winter, or the bees will freeze and die.

Field Bees

Field Bee

Worker bees who fly away from the hive in any role are field bees.

A field honey bee is an adult female bee. It is about the size of a normal paper clip, or just over three quarters of an inch long, or 22 mm. These bees have two pairs of wings and have fine, feather-like hairs that cover their entire bodies. They have a long, tapering tongue that is specially adapted to lap up nectar from flowers. The eyes of a honey bee have multiple lenses that allow them to see a range of colors and patterns, making it easier for them to find flowers with nectar and pollen. The primary function of a field honey bee is to collect nectar, pollen, water, and propolis. Nectar is collected and carried in honey stomachs, and pollen is carried on their hind legs in pollen baskets.

After collecting the nectar and pollen, the field honey bee returns to the hive, where it regurgitates the nectar to receiver bees, who carry it for processing. The pollen, too, is taken, altered, and stored in cells, later to become royal jelly and for nourishment by the bees. As bees tend to the blossoms, they inadvertently transfer pollen from one plant to another, ensuring cross-pollination and genetic diversity within plant populations.

Absconding Bee

Honey bees abandon their home when life is threatened.

Bring in the bee movers; the bees have decided to abandon their hive. Their home, remaining pantry, all they have built, the incubating young, everything they had, is not enough to make them stay. The door is left open without even a goodbye. Colonies pack up and abandon or abscond when something is not suitable for them. Unlike swarming, where only a part of the colony leaves, during absconding, the entire colony leaves, taking provisions for the journey to the new hive. These bees may take resources from their abandoned hive immediately following their departure.

Bees that remain in a hive after the colony swarmed, were either newly emerged bees or those returning after their colony left. The remaining bees missed the boat. Many reasons exist for bees abandoning a hive: resources are poor; too far away from a water source or drought; excessive prolonged smoke in the air and forest fires; disease; overrun with varroa mites or the small hive beetle; attacks from ground crawlers; robbing by nearby bees, hornets, and wasps; threats by rodents; being decimated by grazing animals or bears; and last, abuse or neglect.

Air Defense Bee

Older worker bees patrol and defend the surrounding area around the hive.

All bees buzz, even flies buzz while they fly. Buzzing is what is heard when their wings move rapidly. The difference between bees and other flying insects is bees have four wings, a pair of wings on each side of their thorax. Each pair of wings are uniquely connected with hooks called **hamuli.** These specialized hooks connect a pair of wings together while in flight. A pair of wings fluttering independently have less power to lift the bee, compared to a pair of wings connected acting in unison. Bees can connect and detach the wing hooks prior to flight and after they land.

When a colony is threatened by company outside the hive, older soldier bees fly out to meet the unwanted visitors. The bees have their usual low hum while flying. That hum increases in pitch when they feel threatened, or their wings move at a faster rate. At first, the responder defense bees will warn the visitor by bumping into them, and if the threat remains, the pitch sound of their wings gets higher, and stingers come out. Once a bee stings, the injured bee releases an alarm pheromone. This alarm scent alerts other defense bees, who come to aid in the attack.

Field Bees

Bee-Lining Bee

Bees are not connected to the internet, nor do they use the Earth's magnetic poles to find their home. Bees have an innate ability to know precisely where home is and how to go straight there. This is called a **beeline**. This ability is based more on navigation through landmarks and where the sun is positioned in the sky, like sailors using sextants on a sailing ship. Bees cannot see stars, but they can see poles, buildings, and other landmarks, and this guides them in their journey. Bees can also see the intensity of light through their third eye, called an **ocellus**. They can also see the ultraviolet spectrum of light. So, with these amazing skills, bees fly based on their relationship to the sun during daylight hours and use landmarks to get them home after a long, laborious journey.

A beeline is the straight line from the last foraging location directly to its home hive.

Cleansing Bee

Bees prefer to relieve themselves outside. This happens all year, but poo is more visible on snow.

When around an active hive, watch out for falling bee poop. Bees poop outdoors. During the end of winter, when temperatures rise, bees leave the hive to relieve themselves. At about 50°F (10°C), bees take flight in spring and release the long-held poo immediately after leaving the hive, raining down brown goo out front. Poo spatters are easily visible on white snow as well as on the front of their hive. Multiple reasons exist for spring cleansing flights. Bees defecate all year, but it is more visible in the winter snow. Factors playing into why bees' poo so much in spring can be due to eating unripe stored honey creating a sickness called **Nosema**. Any time a colony is confined and under stress, once released, they will engage in cleansing flights or bee poo raining down surrounding their hive.

Drifter Bee

Bees can drift from colony to colony if the hives are close together.

House numbers on a hive do not help bees find their home. Bees can drift from hive to hive when stacked side-by-side. When bees drift, they inadvertently leave one colony and return with forage to a foreign colony. A returning foraging bee can accidentally enter the wrong nearby hive. Lucky for the absent-minded bee, since she is bringing pollen or nectar with her, she is allowed to enter instead of being viewed as a robber bee. Drifting can become a problem when more bees drift one way, leaving one colony short of bees while the other colony becomes overcrowded. When bees drift from a nearby colony to a weaker colony, it can give them added resources and increase their population. Sharing diseases and invasive pests can also result from drifting.

Flower Pollen Dancer

Bees can be found rolling inside flowers to collect pollen. They push the sticky, wet pollen into their pollen baskets on their hind legs and return to the hive.

Do bees carry baskets on their hind legs? They are called baskets but are actually stiff hairs called **corbiculae**. As a bee flies, it creates static electricity. This is like the electricity created by walking on a carpet while wearing socks. The bee is positively charged when it lands on a flower, and the flower is negatively charged. Scrabbling is a rapid movement over a flower (pollen break dancing) to dislodge and allow pollen collection. This creates an immediate attraction, causing the bee to be covered with wet pollen. The bee then grooms itself, pushing all the pollen onto the stiff hairs on its hind legs. When the pollen baskets are loaded, a beeline trip home is needed, where the receiver bees remove the pollen for processing.

Field Bees

Foraging Bee

Bees who leave for pollen, nectar, water, or propolis ingredients.

A foraging honey bee begins her journey by accepting the call of a shaking bee and leaving the hive in search of forage. These sources can include flowers, overripe fruit, unprotected sweet substances, and even robbing resources from nearby hives. To fill up their expandible internal honey stomach, a bee can visit 1000 flowers. Once back at the hive, the foraging honey bee will transfer the nectar or pollen to the receiver bees for further processing. The foraging bee will then return to the food source to collect more nectar or pollen, repeating the process many times a day. Foraging honey bees also play a significant role in pollination. As they visit flowers in search of nectar, they inadvertently transfer pollen from one flower to another, helping to promote fertilization and the production of fruits and seeds.

Garbage Removal

Bees remove debris from the hive to prevent robbers from being attracted to their hive.

A honey bee hive is well-kept and clean. Risks of bacterial and fungal infections, as well as invasions from predators, can occur when debris is discarded too close to the hive. Mold can form and grow rapidly in the bee's humid environment. Air flow is essential for supporting optimum humidity but is also used to blow unwanted pieces of hive and discarded bee matter out of the hive. This waste can be dead bees or parts of them, larvae or diseased larvae, wax, which is bee fat, and dropped pollen. All these substances are food for other creatures. The beehive is a hidden pantry for nearby insects, birds, and walking animals, big and small. It is ideal for bees to carry their dead carcasses as far as they can get them, so they do not attract attention. If this job is left undone, attacks will soon follow and can overrun a colony. When bees discard waste too close to the hive, it can attract attention to the hive and require the defense bees to ward off attacks.

Field Bees

Guard Bee

Guard bees wait at the front of the hive for a signal to defend. Unrecognized pheromones alert guard bees into action.

Guard bees are at the hive entrance, watching over the comings and goings of every bee that enters or exits. They are the primary agents of hive defense and signify an alarm by hitting and stinging so that sound pitches go up. Guard bees protect the hive from potential threats, both internal and external. If a bee from another colony tries to enter the hive, the guard bees will put up a fierce defense, often resulting in the death of the intruding bee.

But guard bees do not just protect their hives from other bees. They also fend off predators, such as birds, which might look to raid the hive and steal its valuable honey stores. They do this by stinging the attacker repeatedly, sacrificing their own lives if necessary to defend the colony. Guard bees can recognize and respond to individual cues from their own colony, which aids in immediate defense. Guard bees have physical adaptations that make them particularly well-suited to their job. They are expendable, mature, aged worker bees. They are able to fly, increase their buzz pitch as a warning, sting as a last resort, and die outside of the hive where they pose no threat to their colony.

Orientation Bee

Orientation flights gradually increase in radius from the hive, helping bees memorize their location.

Honey bees fly outside the hive to familiarize themselves with their surroundings. Young bees take short flights around the hive to learn the location, landmarks, and smells of their environment. These orientation flights help bees navigate to and from the hive when foraging and find potential nest sites. Orientation flights are taken by young worker bees, imported adopted bees, or after a change in the location of the hive. This first flight lasts from between 10 and 30 minutes, during which they make increasing spiral flights away from the hive. These flights are characterized by their erratic movements, with the bees zigzagging and circling in the air. Through these flights, bees familiarize themselves with the surrounding environment, memorizing landmarks and the hive's location in relation to them.

Packaged Bee

Bees are not arrested and jailed, but packaged, so they can be safely moved to new homes.

Bees are not sentenced to a bee jail. But commercial bee yards weigh and package live bees so they can be moved. These packages may appear to look like a jail, but they are a way to move colonies. These caged packages vary but are usually about 3 lb. (1.361 kgs) or hold about 7500 bees with one newly mated young queen. The young queen is placed into a personal cage with personal attendants who care for her. When the bees were culled from a large colony, it had a queen mother. Once the bees were placed with a new queen, these bees viewed this new queen as a threat for four days. During these introduction days, the worker bees try to kill her. This new queen smells different as she has different pheromones from the one, they knew and loved. After about four days the scent of the old queen dissipates, and they learn the new queen's scent. Once they are familiar with the scent, they consider this new queen to be their own mother. At this point, the bees are no longer trying to kill her but trying to rescue her. This fourth day marks the period when the queen can be released into the cluster of bees to begin laying her own offspring.

Robber Bee

Old foraging bees look for resources. Sometimes the free resources are hard to find, so they turn to robbing for them.

Honey bees primarily feed on nectar; a sugary substance collected from flowers. This nectar undergoes a process of dehydration, turning it into honey while in the hive. This stored honey serves as a food source for the colony during periods of food scarcity. During times of scarcity, bees look for alternative sources of carbohydrates. Bees are a very social insect and visit nearby colonies daily. When these colonies are under-defended or display leaking honey, visiting bees will return with the news to their hive. This results in a robbing frenzy where many bees attack a hive for their military spoils. When these bees are in a frenzy, they are very agitated and are in attack mode. Not just one family, but many clans, as well as wasps and hornets, engage in these frenzies. Robbing occurs with any discarded or unprotected sugar. Hummingbird feeders are a common source for robbery.

Security Enforcer Bee

*All returning bees need the scent of the queen or supplies.
If there is no queen scent and no supplies, then they are treated as robbers.*

Bees are very social insects and visit surrounding colonies daily. They are visiting to assess the defense and what honey stores they have. When these visitors get too close, an enforcer bee comes to see if they are friends or foes. When this visitor is considered a threat, bees can appear to wrestle one-on-one or in a tag-team fight. The enforcer can be seen using their stinger on this unwelcome visitor until it retreats or dies.

Security Examination Bee

Bees regularly check incoming visitors by smell to see if they are foreigners or family. Their queen's scent on them is the password.

Entrance into a beehive is an easy task; does the bee have pheromones matching their queen or not? Exceptions occur when a worker bee drifts to this foreign hive accidentally. If the bee is bringing resources to contribute to the new colony, it is then permitted into the hive. Once permitted into the hive, the bee is adopted into the family. Drones are given an exception, as they are vital in mating with a new queen, when it is needed. If bees smell right to the investigating bee's antennae, then a pat on the back, and in they go. Robber bees are easy to identify; they appear agitated, have an unfamiliar scent, and do not bring anything to contribute. This robber bee is then fought off or killed.

Field Bees

Senior Bee

Old bees naturally fly away from the hive to prevent health and safety risks to the colony.

There are no caskets or cremations for bees when they die, nor are there memorials to remember their amazing achievements. The old bees tend to die isolated from the hive, where they pose no risk to the colony. The bee is focused on the colony's well-being instead of itself. The senior bees may lose the ability to fly by wearing out their wings from the thousands of miles they traveled and the almost endless fanning they performed in the hive. This bee may have met its fate by being caught, consumed, and acting as further nutrition for another creature. It may have lost its way back to the hive and been left alone to die. Many reasons exist for an older bee to die away from the hive. Dying away from the hive saves the bees from having to remove the carcass and the robbing threat it would have been had it stayed.

Scavenger Bee

Bees are the best recyclers of discarded sugar on the planet. From dumpsters to coffee and smoothie shacks, honey bees can smell the treat and are determined to get it and return home.

Honey bees are highly effective at collecting nectar from blooming plants, but there are times when plants are dormant or there is a drought. They are constantly searching for resources to take home for the bees that are waiting for nourishment. People consume vast quantities of sugary sweets and discard their wrappers and containers when done. These discarded wastes consist of sugars that bees collect and can cause environmental harm to nearby insects. Honey bees are opportunists and will collect what is close and easy. Unprotected garbage waste in amusement parks, gas stations, and dumpsters can all take the bees away from pollination for neighborhood garbage removal. Garbage cans and dumpsters can have lids to prevent this robbing, but sugar during a flower famine caused by summer heat exists in the oddest of places.

Field Bees

Scout Bee

Scout bees look for potential homes their colony can swarm or abscond to, as well as resource locations.

Not all scout bees become waggle dancers and shaker bees, sharing messages about resources. Scout bees look for future sites the colony can claim for a future home. Scouts bring messages that other bees will investigate. The scouts determine whether a spot is acceptable and later guide their colony's swarm with its queen to this new home. This specialized bee is a key to the survival of the honey bee colony, allowing for the effective communication of critical information.

Undertaker

Dead bees are removed and discarded far away from the hive, as they are a health and safety risk to the colony.

Worker bees recognize and remove dead bees from the hive using scent identified as cuticular pheromones. This pheromone scent stimulates the need to remove dead bees. Dead bees in a healthy colony are removed by these undertaker bees and, when possible, discarded over 15 feet (4.57 meters) away. Excessive amounts of dead bees can be discarded below the hive entrance. If excessive dead bees are found around the hive, a significant event has occurred related to either predators or pesticides. Under normal conditions, bees are programmed to remove dead bees.

Field Bees

Veteran Bee

When worker bees sting, their barbed stinger is torn from their body, eventually killing them. Even while detached, the stinger keeps pumping venom.

A defense bee losing its stinger sets off an alarm pheromone, alerting nearby defense bees to the threat. The venom a bee injects into the skin is called **apitoxin** and has a smell like bananas. An average hive has less than 20 guard bees on patrol, but this number can increase depending on external issues affecting the colony. This type of pheromone is called a **releaser pheromone**, as it only affects the defense bees nearby and dissipates quickly. The sting may hurt and swell differently depending on the person stung, ranging from a mild itch to a major hospital emergency when an anaphylactic reaction occurs. Once a bee stings and loses its stinger, it only lives for a few more minutes.

Water Bee

Water is vital to a colony, as it is used to reduce heat and increase humidity for the young during the warm season.

Water is one of the four essential resources in a bee's world, alongside nectar (supplying carbohydrates for energy), pollen (offering protein), and propolis (serving as a disinfectant and sealant). Bees need water for supporting humidity levels essential for their young. During the season when eggs and larva are present, bees need a continuous supply of water carried into the hive. This water source needs to be close so they can keep ideal humidity. Once the water is in the hive, they fan it into the air, like a humidifier. Water can be collected from algae-covered sludge ponds to dripping spigots, chlorinated pools to liquid sewage; all types of water can be used equally well. During the winter months, moisture inside the hive can be deadly to bees, as it can hinder the production of heat.

Drone: The Mating Specialist

Drone

The drone is the male bee who is solely responsible for mating with virgin queens and adding diversity to the queen's offspring.

Drone honey bees do not have a father, and as such, they only have one set of chromosomes. This single set of DNA is called **haploid**. Female bees are **diploid**, meaning they have two sets of chromosomes from two separate parents. A drone lives for up to 55 days. Its purpose in life is to wait until a virgin queen needs mating. Drones will fly to a drone congregation area (DCA) to compete against each other for the fatal insemination of the queen. When a drone ejaculates, it is so powerful that its endophallus ruptures, breaks off while still in the queen, and the drone falls to the ground. Drones can also meet their final fate by being a nutritional morsel for predators. Drones have also been known to die of heat-related stress. The drone overheats at temperatures above 107°F (42°C), and as a result, their internal penis ejaculates and exits their abdomen in a convulsive manner.

Band of Brother-Husbands

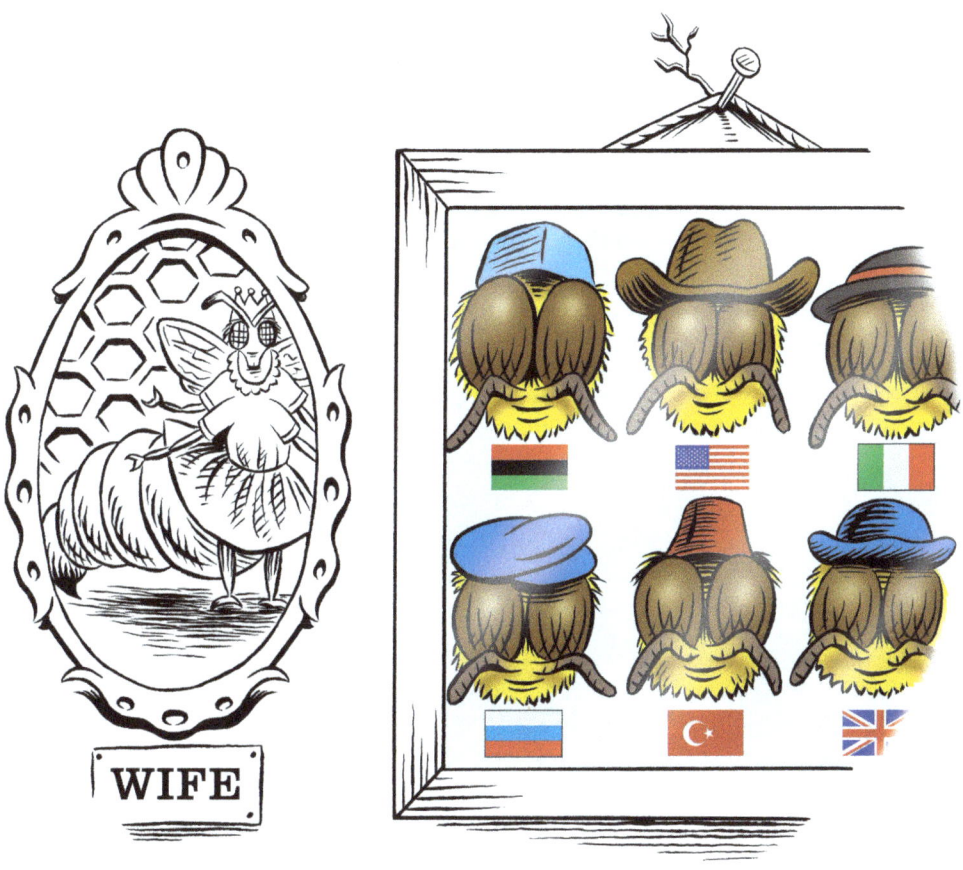

Drones can travel miles in search of a virgin queen. Up to 15 drones can mate in succession with one queen. The genetic diversity of the queen's offspring increases from a well mated queen.

Drones are haploid, having a mother but no father. The primary duty of a drone honey bee is to mate with a queen bee. A queen bee can mate with up to 15 drones. A fully mated queen may produce offspring that appear to be different races of bees due to the race of drone bees she mated with. This task is crucial since the queen bee is solely responsible for laying female eggs in the hive. Once the drone bee mates, it dies, and the drone bee's role ends. All this diversity means that it is indeed possible for different races of honey bees to come from the same queen bee. If a queen bee is mated with drones from another race, it is quite possible that some of her female offspring will display characteristics of that race of bee. Beekeepers often selectively breed their bees to produce desired characteristics, such as higher honey production, resistance to disease, or better temperaments. This can involve choosing certain queens and drones for breeding based on their genetics.

Drone Foster Family

In spring, drone populations increase to ensure virgin queens' mates. Drones may fly miles away in search of a virgin queen, spreading genetic diversity to other areas.

As drones fly farther and farther away from their home hive, they present DNA differing from their own. This DNA gives new traits to the new queen's offspring. The traits can be better or worse than the family of the virgin queen they mate with.

Drone: The Mating Specialist

Drone Fraternity

***All drones serve one colony's need: mate with virgin queens
in their local hangout, the drone congregation area.***

Drones feed each other through trophallaxis feeding and cannot defend themselves. They travel daily to their drone congregation area and return to a hive for nourishment if they survived the ordeal. There is a competitive nature with each other in mating, but besides mating, they can have a good relationship with foreign drones like close brothers.

Drone Grooming

Other than mating, drones spend time grooming themselves.

Initially, when a drone emerges from its birth cell, it grooms and dries from incubation. This grooming continues for the life of the drone. This behavior is not to attract the perfect queen but as a standard life function. A drone has hair to keep clean.

Drone Race Track

When a virgin queen enters the drone congregation area, drones compete for their chance to mate.

Drones emit pheromones, which attract other drones. This attraction helps drones unite at drone congregation areas (DCA). A DCA is a geographical location where drones gather and are typically found in open spaces such as fields, meadows, or clearings. Drones use landmarks such as trees and buildings to navigate to this mating hangout. In the DCA, drones fly in a large oval pattern, waiting for a queen to arrive. When the queen arrives, the drones compete to mate with her, with only a small percentage being successful.

Environmental factors, such as weather and habitat loss, can negatively affect drone congregation areas. Extreme temperatures, rain, or wind can prevent drones from flying to the DCA, while the loss of open spaces and landmarks can disrupt the navigation of the drones toward the congregation area. The decline in the number of drone congregation areas can have severe implications for honey bee reproduction and the health of the honey bee population.

Food Beggar Drone

*While the drone is emerging from its birth cell, it begs for food.
Even as adults, drones seek food found by worker bees.*

When a drone emerges from its birth cell, it is fed for the first seven days of its life. A drone does not know how to forage. After the first seven days, it looks for food already prepared. Once a drone has matured, it can engage in mating. Weak, unhealthy, starved drones are unfit for breeding. Allowing the drones access to food stores gives the colony insurance for the day when the queen needs to be fertilized. Well-fed drones equate to successful future queens, and healthier worker bees. Available prepared food is key for a drone to be able to perform this insect consummation.

Drone: The Mating Specialist

Freezing Drone

To protect essential honey reserves, worker bees exile all drones from the hive in the fall. These drones die of hypothermia or starvation.

Drones are allowed to eat and roam the hive managed by female worker bees until the end of the foraging season and temperatures drop. At winter's door, the resources in the hive become sacred to the generous worker bees, and they expel all the drones to meet their fate outdoors.

Predator Fodder

Drones are larger and slower than worker bees, making them easier to see and capture by predators.

Drones are larger and have wider bodies than worker bees, which may make them easier targets for predators like birds and other insects. During these times, drones will fly out of the hive and congregate in a specific area to wait for a queen to fly by. This makes drones highly visible and vulnerable to predators during these specific times. On the other hand, worker bees are constantly leaving and returning to the hive to collect pollen and nectar, which makes them a more consistent target for predators. However, worker bees have evolved various strategies to protect themselves and the hive, including stinging predators and working together to fend off attacks. Drones tend to fly solo or in small groups in search of a queen to mate with. This behavior makes them more vulnerable to predators, as they are not protected by the collective defense mechanisms of the colony. Additionally, drones do not have stingers and are unable to defend themselves against predators.

Activities & Famous Bees

City Bee

Residential bees pollinate multi-floral plants in gardens, parking strips, and parks. When forage is lacking, bees recycle discarded human trash sugars.

City bees do not drive cars or wear fine clothes; rather, they live among us, using what we have planted or discarded. They can pollinate fruit trees and garden plants, increasing the eventual harvest. They can also visit trash containers to retrieve discarded, coveted sugars. The biggest challenge in the city is overcrowding, or too many colonies to support an area. Bees expand and grow during times of bloom, but the sad reality is that when the blooms are gone, the bursting colonies no longer have enough blooming resources to sustain them through the remaining year and future winters. A succession of blooming plants in the city would be ideal, as it is in undisturbed areas in nature. City folk also add issues by using pesticides without reading the directions, inadvertently creating threats to all nearby bee colonies and other pollinating insects. Flowering plants, like azalea and rhododendron, can make bees sick, and the honey can be harmful to those who eat it. Not all blooming flowers supply nectar or pollen for insect pollinators. Flowers that do not produce nectar, pollen, or produce a scent are called **anemophilous** flowers. These flowers are primarily wind and rain pollinated.

Fossilized Bee

Ancient bees have been found in the fossil record, preserved in rock and amber.

Wasps are evolutionary cousins of the bee. When the wasp changed its diet from carnivorous to vegetarian, it became the first bee. Preserved insects have been found in amber and fossilized tree resin. Bees in amber are rare; ants seem to have become trapped in tree resin the most. The most famous bee fossils discovered come from the Messel oil shale in Germany, dating to the early Eocene period (approximately 56.5 to 35.4 million years ago). These early finds show bees existed around 50 million years ago. These fossils, identified as *Apis nearctica* and *Apis lithohermaea*, look like modern-day honey bees with minor variations.

In the United States, a fossil of an extinct honey bee was found in Nevada, dating back to the Miocene period, 14 million years ago. This fossil from the western United States shows honey bees were native to this region anciently. The La Brea Tar Pits in Los Angeles, California, have been discovering fossils from the period 40,000 to 8,000 years ago. Most of the unlucky creatures found are large extinct mammals, but they are now discovering insects of all sizes. Perhaps one day an unlucky honey bee will be identified there and instantly become famous.

Garden of Eden Bee

Fruit trees and blossoming plants in the Garden of Eden had honey bee pollinators.

There are no early biblical references saying that honey bees existed in the garden paradise in the east, in Eden. In Genesis 2:19, God created all the animals and creeping things to help Adam. He was then given the earliest recorded task of naming them all. This Genesis account in 2:9 also says that all kinds of trees were planted to provide food. One tiny insect created to help Adam was the honey bee. Could Adam be the first beekeeper? Bees pollinate trees and blossoms, essential food for this first vegan biblical family. However, once expelled from Eden, their diet evolved to include meat.

The honey bees' presence is also found in other recorded Bible passages, such as Land of Milk and Honey, and was used to describe the Promised Land of Israel. The use of honey and its connection to deity continues through the Quran. The Quran refers to paradise as a place with rivers of honey flowing in Jannah, with Jannah being the final location of the righteous.

Allah is also believed to have ordered the bees to make honey, which is to be used to heal the human body. Mohammad counseled a man who had stomach problems to "Let him drink honey." Once the council was followed, the brother was cured. It was believed that thousands of remedies entered the stomach when honey was eaten. Later in the biblical account, Matthew 3:4, it must be mentioned that John the Baptist, while in the wilderness, ate locusts and honey. The Bible and the Quran both equate honey and bees with heaven and places of paradisiacal beauty and health.

Pilgrim Bee

European honey bees migrated to the Americas on Puritan colonists' ships.

All aboard! Even the migrating honey bees set down in the ship's hull. The exact moment European honey bees came to the new world is uncertain, as colonists arrived with undetailed ship cargoes. The *Discovery* was the first ship known to be carrying European honey bees to America.

The Council of Virginia Company in London wrote in old English on December 5th, 1621, "We have by this ship … sent you divers sorte of seed and fruit trees, as also Pidgeons, connies, Peacocks and beehives …" (Crane, 1999). It is possible that more bees could have arrived prior, but they definitely continued to arrive in North America after that date. Once bees arrived, swarms aided their colonization of the new world. As the colonists migrated west, and around the tip of South America to the west coast of the United States, so did honey bees.

Fossil evidence now shows honey bees were native in the Americas 14 million years ago. Christopher Columbus is also known to have transported honey bees on his voyages, and it is possible that on these voyages, bees may have escaped and established colonies in the Americas. Stingless bees of the Amazon have also been known to produce lesser amounts of honey and wax. Artifacts made of beeswax were discovered in the ancient Mayan ruins, dating back to as early as AD 700. Honey was also used by the Mayans and given as payment of taxes to the Gods.

It is thought that the Pilgrims brought honey bees to America for three primary reasons: to harvest the sweet by-product of honey, to produce beeswax, and for the essential role they played in their crop pollination. In Europe, honey was a valuable commodity used for a wide range of purposes, including food, medicine, and cosmetics. Colonists saw the advantages of bringing honey bees to the Americas, as they were familiar with how to manage bees for crop pollination and needed the honey and wax from the hive for everyday use.

Weaponized Bee

Active honey bee hives have been used throughout history in warfare. Historic siege warfare dating back to the Romans used honey bees and their honey as a tool of war. Bees have been dropped from castle walls onto invading armies, as well as tossed over walls and into tunnels to clear out opposing troops. Honey bees marched against the Muslims during the Third Crusade, under the employ of King Richard. Poisonous honey was used to debilitate the Roman army. Pirates used honey bee hives dropped onto ships to clear off the decks prior to boarding. In 946 BC, St. Olga used fermented honey, called mead, as a gift to intoxicate herself prior to an attack. Nuns toppled hives against plundering intruders.

So effective were bees in protecting property that castle walls were made with bee holes to hold hives. These would prevent scaling the walls and could be used to toss at the intruders. During the American Civil War, artillery from the South hit an apiary as the Union troops were passing by, giving the Southern troops a first advantage. Insects have been used in warfare almost continuously since the progressive Julian calendar began in 46 BC. Some insects were used to distribute venom, carry diseases, or send other pathogens, later defined as entomological warfare. This type of warfare was such a global threat that, in 1972, under the Biological and Toxic Weapons Convention, the use of insects to administer agents or toxins for warfare was against international law.

Throughout history, honey bees were used in warfare to sting the opposing army. Rare plants produce deadly honey, used as bait for soldiers to make them sick or weak.

Activities & Famous Bees

White House Bee

President Washington and President Thomas Jefferson kept bees on their estates, but First Lady Michelle Obama was the first to bring bees to the White House.

Presidents Washington and Jefferson both had honey bees, but the White House's first exposure to on-site honey bees pollinating the gardens was during the Barack Obama presidency. First Lady Michelle Obama was the first person to encourage and bring bees to the White House. As an organic gardener, Michelle needed bees for pollination. Charlie Brant was employed to manage the apiary and has now been the White House beekeeper for more than two decades.

Wild Feral Bee

Honey bees have lived on their own in nature's cavities since dinosaurs walked the Earth. Today, honey bees seem to be domesticated, but we have learned how to utilize their natural activities for our benefits. People have added a hive that can be used for migratory beekeeping, have learned how to harvest products bees make, have developed removable frames for health inspections, and have learned how to reproduce worker bees and queens as needed. In the end, all honey bees are wild bees, and are led to do what their DNA instructs them to do. Beekeepers have learned to use honey bees by learning how they act and multiply in artificial, manufactured hives. These modern-styled hives may appear to have domesticated the useful insect, but bees still function as if they were still in a feral hive they chose in nature. It is common for bees to swarm or abscond from an apiary to find a new home. These new homes can be hollow trees, buildings with holes in them, ground cavities, caves, or even open-air hives built from tree limbs and house soffits. Some bees have lived for many years on their own, under the noses of people living nearby. These feral bees pose threats due to the risk of disease and the spread of Africanized colonies that overtake traditional colonies.

Bees living in unmanaged hives without the care of a beekeeper are called feral bees.

Closing Remarks

I hope that through reading this book you have gained insights into the various types of roles bees perform, and how bees communicate with each other and the world around them. Bees communicate and engage in activities from when they first emerge from their birth cells to the time they die. No memorials exist for bees once they are gone, yet they served their colony family selflessly their entire lives.

Bees also provide us examples through their dynamic lives how we could improve life within our human colony.

It has been my great pleasure writing this book for you to enjoy.

With thanks,
Albert B. Chubak

Printed in the USA
CPSIA information can be obtained
at www.ICGtesting.com
LVHW071159040424
776423LV00055B/1090

9 781038 300362